The Hierarchy of Energy Systems

From Atom to Society

AF740396

SYSTEMS SCIENCE AND WORLD ORDER LIBRARY
General Editor: Ervin Laszlo

Explorations of World Order

DE ROUGEMONT, D.
The Future is Within Us

DILLOWAY, A. J.
Is World Order Evolving? An Adventure into Human Potential

GIARINI, O. & LOUBERGE, H.
The Diminishing Returns of Technology: an Essay on the Crisis in Economic Growth

LASZLO, E.
The Inner Limits of Mankind: Heretical Reflections on Today's Values, Cultures and Politics

LASZLO, E. & BIERMAN, J.
Goals in a Global Community
Vol. 1: Studies on the Conceptual Foundations
Vol. 2: The International Values and Goals Studies

LASZLO, E. & VITANYI, I.
European Culture and World Development: UNESCO Joint Studies for the European Cultural Forum

MARKLEY, O. & HARMAN, W.
Changing Images of Man

MASINI, E.
Visions of Desirable Societies

SAUVANT, K.
Changing Priorities on the International Agenda: The New International Economic Order

TEVOEDJRE, A.
Poverty: Wealth of Mankind

Innovations in Systems Science

AULIN, A.
The Cybernetic Laws of Social Progress

COOK, N.
Stability and Flexibility: An Analysis of Natural Systems

CURTIS, R. K.
Evolution or Extinction: The Choice Before Us

GEYER, R. F.
Alienation Theories: A General Systems Approach

GEYER, R. F. & VAN DER ZOUWEN, J.
Dependence & Inequality: A Systems Approach to the Problems of Mexico & Other Developing Countries

JANTSCH, E.
The Self Organizing Universe: Scientific and Human Implications of the Emerging Paradigm of Evolution

NOTICE TO READERS

If your library is not already a standing order customer to this series, may we recommend that you place a standing or continuation order to receive immediately upon publication all new volumes published in this valuable series. Should you find these volumes no longer serve your needs your order can be cancelled at any time without notice.

1800107826 TELEPEN
QC73.F2
UCW LLYFRGELL LIBRARY ABERYSTWYTH
DILEWYD O STOC
WITHDRAWN FROM STOCK

The Hierarchy of Energy Systems

From Atom to Society

by

GERRIT B. FEEKES

Research Fellow, University of Technology
Eindhoven, Netherlands

PERGAMON PRESS

OXFORD · NEW YORK · BEIJING · FRANKFURT
SÃO PAULO · SYDNEY · TOKYO · TORONTO

U.K.	Pergamon Press, Headington Hill Hall, Oxford OX3 0BW, England
U.S.A.	Pergamon Press, Maxwell House, Fairview Park, Elmsford, New York 10523, U.S.A.
PEOPLE'S REPUBLIC OF CHINA	Pergamon Press, Qianmen Hotel, Beijing, People's Republic of China
FEDERAL REPUBLIC OF GERMANY	Pergamon Press, Hammerweg 6, D-6242 Kronberg, Federal Republic of Germany
BRAZIL	Pergamon Editora, Rua Eça de Queiros, 346, CEP 04011, São Paulo, Brazil
AUSTRALIA	Pergamon Press Australia, P.O. Box 544, Potts Point, N.S.W. 2011, Australia
JAPAN	Pergamon Press, 8th Floor, Matsuoka Central Building, 1-7-1 Nishishinjuku, Shinjuku-ku, Tokyo 160, Japan
CANADA	Pergamon Press Canada, Suite 104, 150 Consumers Road, Willowdale, Ontario M2J 1P9, Canada

Copyright © 1986 Gerrit B. Feekes

All Rights Reserved. No part of this publication may be reproduced, stored in a retrieval system or transmitted in any form or by any means: electronic, electrostatic, magnetic tape, mechanical, photocopying, recording or otherwise, without permission in writing from the publishers.

First edition 1986

Library of Congress Cataloging-in-Publication Data

Feekes, Gerrit B.
The hierarchy of energy systems.
(Systems science and world order library.
Innovations in systems science)
Bibliography: p.
1. System theory. 2. Social systems. I. Title.
II. Title: Energy systems. III. Series.
Q295.F44 1986 003 86-16904

British Library Cataloguing in Publication Data

Feekes, Gerrit B.
The hierarchy of energy systems: from
atom to society.——(Systems science and
world order library. Innovations in systems science)
1. Force and energy I. Title II. Series
531′.6 QC73

ISBN 0-08-033901-8

Coleg Prifysgol Cymru
The University College of Wales
Aberystwyth
Acc. No. 1800107826
Class No. QC 73 F2

Printed in Great Britain by A. Wheaton & Co. Ltd., Exeter

To Cynthia
and our sons
Fons, Erik and Robert

Contents

1.

Introduction

In Greece at the time of the seven sages, around 600 B.C., the people turned from mythology and the mythical nights turned into scientific days. Myths are like dreams that disappear at daybreak when the Sun rises and dispels the darkness. However, the Sun's rays take some time to break through the shadows of mythical nights, before the world becomes a brighter place.

The word "energy" is ascribed to Heraclitus who used it first; it is derived from *en-ergon* which means "the source of activity". In his book about nature, he wrote: "En-ergon is the father of everything, king of all things and, out of it, all forms of contrast originate. Since 'en-ergon' is common to everything, it is vital for life itself." During the centuries that followed, many attempts were made to define the essence of energy.

1.1. The Development of Energetica

The word energy appeared in the West during the seventeenth century, at the time of a sudden surge forward in the development of mathematics and the natural sciences. In contrast to earlier social thinkers, the new "shining lights" like Hobbes, Spinoza, Descartes, Weigel, Leibnitz and others were not anthropocentrists. They tried to make common beliefs of the day more rational and objective; and their new way of thinking shone like a star that illuminated technical aspects of life.

In the eighteenth century, this way of thinking was strengthened by Fourier, Saint-Simon, Comte and Quetelet. Then, in the second half of the nineteenth century, there was a transition in thinking from mechanistic to energy-oriented. The mechanistic view was based upon the knowledge of mechanics; however, after Joule and Mayer, the many forms of energy were investigated in depth. It was agreed that mechanical energy was just one of these forms.

The Energetica School (up to 1900) comprised the writers L. Solvay, W. Ostwald, W. Bechtereff, T. N. Carver and L. Winiarsky. After Solvay, it was believed that social phenomena were combinations of three factors — organic, inorganic and physical. They saw life as a metamorphosis of energy, and it followed that social differences were energy manifestations.

U.C.W. ABERYSTWYTH LIBRARY

Ostwald had a somewhat more cautious opinion, although subsequent doctrines used energy as a basis for defining some fundamental principles of the social sciences. He said plainly: "Alles Geschehen ist in letzter Instanz nichts als eine Veränderung der Energie." Next, Carver proposed that the life of an individual and the development of people into groups sprang from the transformation of radiant energy from the Sun into human energies. He endeavoured to support this thesis with an analysis of economic phenomena.

Winiarsky came to the conclusion that all physical phenomena were modifications of biological energy which, in turn, was a form of physio-chemical energy.

In his *Kollektivnaija Reflexologia* he summarised 23 laws of physics and biology; he also analysed social phenomena in order to support their similar derivation.

Criticism of these works soon appeared, particularly because the facts used to derive the analogy were a little doubtful. The conclusions were not fully substantiated, since there were several uncertain areas in physics and the social sciences could not boast of much numerical evidence. In fact, these hypotheses came before their time, because the available scientific knowledge was not clearly defined then.

1.2. New Developments

(a) In physics, the fundamental scientific knowledge soon became more clear, being explained in the work of Bachelard, *L'energie est le support de tout.* Additionally, Heisenberg, Paynter, Karnopp, Thoma and Polder had contributed to the development of a general energy theory that led to the general energy system. It was assumed that all forms of physical power could be expressed as products of force and action, i.e. power was the use of energy per unit of time. It followed that some studies in the field of social studies were based upon this revised theory of energy comprehension.

(b) In sociology, the following viewpoint is quoted from *The Energetic Basis of the Social Learning Process* by the sociologist Dr. A. Cornelis:

> "When we speak of a philosophy of life we mean metaphysics in a scientific sense. The notion of energy incorporates here substance, that is the continual component, out of which everything has been constructed that we recognize as structures, as well as the dynamic source that brings about changes and structures."

(c) Social power was the subject of an analysis by Richard Adams, and he made this the basis for a revised energy theory. He started with a system comprising the Sun, Earth and Society and developed it into an energy complex. In support, he gave examples from a wide range of social opinion, from primitive beginnings through to the advanced social energy viewpoints of our time.

On the grounds of this theory, the idea of an international balance of power comprised of supranational blocks was developed. These blocks included the Orient, with China dominant; Latin America centred upon Brazil, North America; Western Europe; Eastern Europe; the Middle East; Africa and Oceania.

(d) Synergetica, or the study of synergism, was an important development which was described by Hans Dekker. In physiology, according to him, synergism, or the co-operation between muscles and muscle groups, had been discussed for some time in connection with breathing and the co-ordination of muscles for bodily movements. In physics, the term synergetica had been introduced by Professor Hermann Haken of Stuttgart, as the name to describe his studies for linking physics with the other sciences, like chemistry, biology and sociology. This confirmed that many systems which appeared to be different on first sight actually displayed a remarkable similarity on further examination.

Now, statistical analysis played an important role in the development of new theories. The Law of Probability, which governs statistics, explains the degree to which properties of a system are influenced by individual components. Their "microscopic" interactions play a part in the overall character of the system. Much is said about the co-operative elements in a complex system that is composed of many smaller subsystems; for example, the atom, brain cells, or individual persons in society. The individuals are united into a system according to an organised arrangement.

The use of statistical methods in physics, chemistry, or sociology was not only possible, but essential for their further development. A detailed "microscopic" examination of each subject was a practical impossibility, nor was it desirable, just as the examination of a living room would be too complicated if it took into account the precise positions and speeds of all 10^{26} atoms in its construction. However, the average speed of the atoms plays an important part in determining the "macroscopic" property that we call temperature.

A single picture from physics will illustrate the effect of microscopic elements on a macroscopic system, namely the transition from electricity to magnetism shown in Fig. 1.

This phenomenon teaches us that a magnet loses its magnetic field when the temperature T rises above a certain critical level, T_c (the critical temperature) and vice versa.

On the magnitude of atoms, Figure 1 explains the reaction, showing that an iron magnet is composed of very many tiny magnetic elements which are called "spins". Spatially, they appear as a disorderly array when $T > T_c$ until $T < T_c$. At the critical temperature (called the Curie temperature), there appears to be a change of phase. The spins show an ordered arrangement like a web of elementary forces (in this example, the exchange medium is the coulomb).

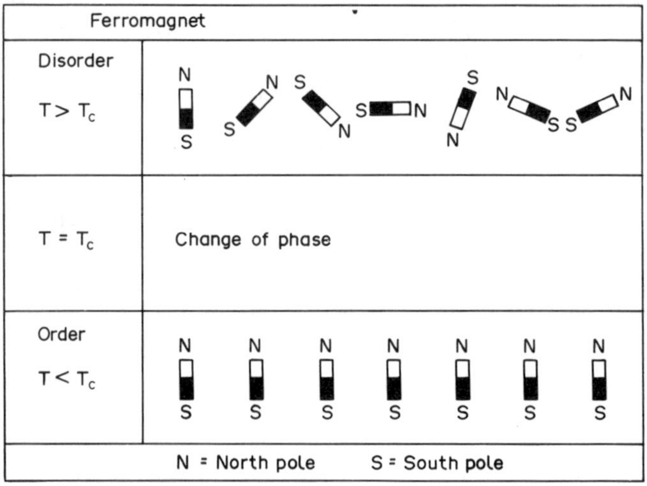

FIG. 1. Transition from electricity to magnetism.

In combination, these minute forces define the magnetic field and show the general directions of its forces. In actual fact, the disorderly behaviour of the spins thermally is well organised magnetically. Above T_c, disorder reigns and the system is unpredictable, but below T_c order takes over and the system is well organised. This struggle between unpredictable and systematic forces was described by Monod as being between "hazard and necessity", and he suggested that each individual element had a fundamental interest in the conduct of the system as a whole. In the case of the iron magnet, when $T > T_c$, the elementary magnetic fields opposed each other, but for $T < T_c$, each minute force contributed to a united force field or system.

When it is realised that there are around 10^{24} spins per cubic centimetre of a magnet, it is absolutely impossible to take into account the effect of an individual spin on the complete system. Under these circumstances, the sensible thing to do is to ignore the superfluous specific elements and to explain the movement of the combined forces in a magnetic field. The spins are minute finite electrical phenomena that react in unison to produce electrical power which is the product of potential difference (force) and electrical flow (action).

It can be shown that magnetism is part of a more complex system in which magnetic power is described again as the product of a magnetic tension (force) and a magnetic flow (action).

1.3. The Universal Energy Picture

Dimensional analysis has many applications for explaining physical phenomena and in subsequent chapters of this book it will be used to construct "A Hierarchy of Energy Systems" (see page 79). In this way, the universal energy picture will be unravelled, commencing with "the big bang" theory, encompassing diverse forms of energy related to life, before arriving at a homogeneous universal system in the centre of which we find ourselves.

Attempts will be made to include the whole of science in this synthesis, with reference to K. E. Boulding's *General Systems Theory*; B. M. Gross's *The State of the Nation* and J. M. Miller's *Living Systems*.

Including these references increases our inventory of knowledge of systems and shows how the different disciplines can be co-ordinated for more general investigations. A "system" is a collection of mutually related components that display a well-structured unity. Some examples of typical systems are:

— the solar system,
— a plant,
— animate life,
— human society,
— a business enterprise,
— an automobile.

When defining a system, a first distinction is in structure and function. In social systems, this distinction is clearly illustrated by Merton and Gross; they showed that the structure of a system was characterised by durability, whilst function was temporal. Miller described the structure of a cell as being analogous with those of organs, organisms, groups, organisations and supranational systems, in which he identified 19 common components. This hierarchy is presented in Table 8, starting with the recording of energy and working through to the differentiation of products.

Under function, the following aspects need to be understood: input, flow and output; the cybernetic principle; the power of systems and energy utilisation; efficiency; etc. In this treatise we will consider only power and energy.

The notion of power can be illustrated with an example of two stone blocks of different weights. Then the lighter block will be moved a greater distance by the same force, all things being equal. The functional components can be expressed in mechanical terms, namely, $M.L^2.T^{-3}$, or power is energy per time. It is known that all conventional forms of energy that are found on the Earth originate from the Sun. Initially, plant systems convert the radiant energy of sunshine into chemical forms like wood and charcoal which release heat when they are burnt.

People have learnt how to convert this heat energy into electricity; in fact, Leibnitz suggested in 1700 that there was a series of energy transformations from plants to people via foodstuffs. Now physics has developed an isomorphic model to explain the relationships between the different forms of energy. Moreover, trends can be shown that this model is also relevant to the social sciences.

The overall picture of the energy forms that exist in our environment is based upon more than conjecture and Chapters 2 and 3 give a survey of the uses for energy and its conversion. Chapter 4 deals with the development of the general energy theory. The possible expansion of this paradigm to include socio-psychological systems will be discussed in Chapters 5, 6, 7 and 8, with the fundamentals of industrial psychology, group dynamics and macro-economics. Chapter 9 describes an actual situation in which human, group, management and economic powers are recognised.

Chapter 10 presents an experiment in order to show the possibility of incorporating all forms of energy together. Subsequently, an attempt is made to solve the problem of identification which bothers people of the present-day "technological society".

2.

Energy Supplies for Life

"Life shows, in spite of its macroscopic diversity, a great unity at the microscopic level" (von Humboldt).

2.1. Unity of Life

Life on earth shows such a bewildering diversity of plants and animals that it is not surprising to learn of an estimate that suggests there are nearly a million kinds of plants and about two million different kinds of animals. Both the plant and the animal kingdoms have an evolution hierarchy, ranging from primitive to highly developed species. They are classified into classes, orders, families, and species. One hundred years ago, Darwin postulated his Theory of Evolution and since then it has been realised that the different forms of life are not invariable and that all higher forms evolve from lower ones by a long development process. The Evolution Theory paints a perfect picture of reality which is based upon undisputable evidence that no serious-minded person can doubt. In fact, it is reasonable to believe that all aspects of life evolved in a similar manner.

Despite its macroscopic diversity (an elephant bears no resemblance to a stinging nettle), life manifests a *great uniformity* at the microscopic and sub-microscopic levels. Each organism is composed of cells which are very similar in structure and behaviour and they are made of the same organic materials. However, when they are united into a single body it bears little resemblance to the features of other composite bodies.

The organic chemicals that comprise a cell are carbohydrates and fats, which are important suppliers of energy. An organic phosphate is known to be the accumulator of energy in cells and without a continuous supply of energy from outside no organism can stay alive.

The cells of an organism need energy for growth, by making larger molecules out of small ones, e.g. protein and amino-acids. In general terms, this energy can be called molecular or chemical energy. As the complexity of an organism advances, its need for energy increases and we need energy to do manual work with the aid of our muscles. Energy is needed throughout our waking lives as well as when we are asleep, because breathing and heartbeats are vital to our survival. All animals need a continuous supply of energy for their bodily functions; to keep them warm and to convey messages throughout their nervous systems. These nervous impulses take the form of minute electrical impulses.

According to the Law of Energy Conservation, the consumption of energy has to be replaced by the production of energy and it is generally known that this requires oxidation. The supply of oxygen is obtained by breathing in air before it is conveyed around the body by the blood whilst the matter to be oxidised is supplied by foodstuffs. These foodstuffs are digested in order to extract the carbohydrates and fats which are conveyed all over the body by the blood too. Some of the food is used for growth by the multiplication of cells and some for the repair of tissues. Superfluous amounts can be stored in the body.

Whenever energy is used, supplies of food and oxygen must be available for replacing it, but food is needed also for providing protein which is necessary for making new cells. Carnivorous animals obtain their food supplies by eating the meat of other animals, herbivores by eating plants only and omnivores (including rats and human beings) eat anything. Finally, parasites are found in both the plants and the animal world. We know that they live on the food eaten by other plants or animals.

Ultimately, plants are the only organisms that require no other life-form as a source of oxidisable matter in order to obtain their supply of energy. With the aid of radiant energy from the Sun, plants produce carbohydrates from CO_2 and H_2O. About half of the energy that they receive is needed for their vital functions, one of the most important being to make protein. Land plants must grow in fertile soil for their supplies of nitrogen which is absorbed by their roots from the nitrates in the soil along with other salts. Other salts that are necessary for plant life are sulphates, phosphates and the elements sodium, potassium, magnesium, calcium and iron.

Rain is essential for land plants, not only as a source of water for photosynthesis in the production of carbohydrates, but for the transport of soluble foodstuffs to the different parts of the plants.

2.2. Photosynthesis and Breathing

Plants and algae (many of which are found in enormous quantities as plankton in the water of the oceans) need only sunlight, air, water and certain soluble salts for their growth. From these things they can synthesise everything that they require, and the energy for this activity comes from the Sun. Without sunlight, there would be no photosynthesis and without photosynthesis the plants could not exist and provide food for animals. *In a nutshell, all life on Earth originates from the Sun.*

Roughly speaking, the chemical processes of life on Earth can be reduced to photosynthesis and breathing. Photosynthesis is the reduction process that produces glucose and absorbs energy. In symbolic notation, the formula that expresses this process is:

$$\text{radiant energy} + 6H_2O + 6CO_2 = C_6H_{12}O_6 + 6O_2 \qquad (1)$$

The total amount of organic matter formed annually by this process is enormous, it corresponds in magnitude to a quantity of carbon equivalent to 10^{14}kg, yet less than 1% of the radiant energy falling on the Earth is used.

In the broadest sense of the word, breathing is a process of releasing energy. In notational form, the formula for oxidation is:

$$C_6H_{12}O_6 + 6O_2 \rightarrow 6H_2O + 6CO_2 + \text{energy} \qquad (2)$$

It can be seen that oxidation is the reverse of photosynthesis, although the original energy was radiant energy from the Sun, the energy given by oxidation may be in the form of chemical, electrical, mechanical or thermodynamic energy. Looking back over these processes, it can be seen that they comprise a life cycle of sunshine + carbon + oxygen. In that part of the cycle where carbon combines with sunshine (photosynthesis), and oxygen is produced, plants and bacteria play an important role. However, where the production of CO_2 is concerned, both plants and animals are important. They are complementary to each other, because animals need the O_2 produced by plants and plants need the CO_2 produced by animals. During daytime, the production of O_2 dominates the cycle, whilst at night the production of CO_2 predominates.

Through the experiments of Priestley, this cycle was known in the eighteenth century, although the gases O_2 and CO_2 were unknown. Priestley demonstrated that a mouse would die in an airtight box, even when plenty of food and water were available. When a plant was introduced into the box, the mouse could live normally.

This observation was insufficient to draw people's attention to the essential fact that the whole biosphere (the relatively thin layer of water, land and air in which life is found) was the medium for this cycle. All organisms in the biosphere are more or less dependent upon each other for their lives.

In the beginning, the Sun's radiant energy is absorbed by plants, but as the cycle proceeds it engulfs everything and brings warmth to life itself.

The use that living things make of sunshine resembles the use people make of the potential energy in a waterfall. As the water falls, all its energy is converted directly into heat, but with the aid of a turbine generator, it can be converted into electricity for manifold uses. Eventually, the potential energy is converted into heat, although along the line it may have provided light for houses and streets, power for railway engines, machinery, computers, medical equipment, telephones, radio and television, washing machines, refrigerators and other modern equipment.

2.3. Energy from Foodstuffs

The reversible equations (1) and (2) show the results of processes that were expressed as photosynthesis and breathing in the last section. In practice, these reactions do not occur directly and intermediate reactions play related parts.

In living cells, this has been solved by the construction of power stations, which are called mitochondria. They are so minute that they are only one-hundredth of a cell diameter and an average cell measures only 100 μm across.

The mitochondria appear to consume glucose when it is not required directly by the organism for its metabolism. As far as it is known, each form of life can convert its food energy into heat energy for the purpose of building up macromolecules from smaller units like proteins and amino-acids. An example is needed to show the inter-relationships between the different energy forms and their relevant values.

The "isotherm" for foodstuffs in living cells shows that there are many more points of agreement between their behaviour and that of fuel cells and with conventional power stations. The mitochondria control the resources of chemical energy in quite small amounts and convey it to special molecules which function as "freight carriers" for transporting energy to those places where it is necessary to perform natural activities for life. It is better to talk about "free energy" here; that is, energy which is not reduced to heat but is freely transferred to places where any form of energy is needed.

The oxidation of a mole of glucose at a pressure of one atmosphere and temperature of 37°C can be expressed by an equation:

$$C_6H_{12}O_6 + 6O_2 = 6H_2O + 6CO_2$$

This reaction produces heat too, but the mitochondria prevent the oxygen from contacting the electrons; however, electrons seem to flow in a more discontinuous manner from the anode to the cathode than in a fuel cell.

They are assigned to coenzymes and cytochromes which function as carriers, forming an integrated system so that each carrier becomes more electronegative and has a greater affinity for electrons. The electrons are passed from carrier to carrier in a chain in which each link acts as a "giver" and "receiver" alternately. Finally, the electrons and hydrogen-ions react with the enzyme, cytochrome oxidase, which activates the O_2 and hydrogen-ions to combine and form water.

2.4. The Assignment and Storage of Free Energy (the Role of Phosphates)

After oxidation, a potential difference in the mitochondria exists, due to the movement of electrons. At least three of these movements show a decrease in free energy which is large enough to be useful for supporting life in a unique way. It is not certain how the electron movements behave chemically, but the ADP changes into ATP. This change results in an increase of free energy with an efficiency of 45%, which is better than a modern steam engine. The change can occur everywhere that it is necessary for the function of an organism; through hydrolysis the reverse action is possible, and again free energy is liberated. The free energy changes to other forms of chemical energy by an increase of macromolecules (biopolymers); or for doing work by contracting the muscles; or for propelling electrical impulses along the nerves. The conversion of the energy released by the oxidation of ATP is analogous to a rock sliding down a hillside. Sliding freely, like free combustion, eventually transforms the potential energy into heat energy.

It can be imagined, however, that the progress is stopped by a spring that can be kept in a tense position by a catch. Then the rock slides a little further and winds up another spring, etc. Thus, the potential energy can be used again and again.

The function of ATP is one of the most universal characteristics of life. It is responsible for the storage and assignment of free energy in all living organisms. *ATP is the universal currency of the energy market of cells and it proclaims the unity of life on Earth.* No single organism can function without making use of the potential energy of ATP, and it is probable that it played an important part in the origin of life.

2.5. The Role of Water

Generally speaking, oxygen is released in photosynthesis, as shown in equation (1), and most of the oxygen in the Earth's atmosphere originated in this way. It is debatable whether this oxygen came from CO_2, or H_2O, or from both. In the reverse equation (2), the oxygen of the air produces H_2O, whilst the oxygen in CO_2 originates from glucose or other foodstuffs and water.

A similar conclusion can be drawn for photosynthesis and for breathing, that is that the oxygen released came from H_2O rather than from CO_2. It is obvious[1] that water flows between two points when there is a difference in pressure. So it is in plants; when water flows from the roots to the leaves,

[1] Th. Alberda *et al.*, *De Groene aarde; biologie en ecologie van de plant.*

the pressure must be greatest in the roots. Actually, it would be more appropriate to say that suction pressure in the roots is least.

Also, the movement of water from soil through plants and into the atmosphere has to be taken into account. In the following diagram the suction gradient is shown in five stages.

Atmosphere and air pockets around the leaves	I	↑
Breathing and intercellular transport through the leaf	II	
Wood vessels in the leaf and stem	III	
Wood vessels in the root and its parenchyma	IV	
In the basic layers of the root tissues	V	

This gradient can be demonstrated by the connection of manometers to different parts of a tree. The difference in suction between two points will be given by $Z_2 - Z_1$, where Z_2 denotes the greatest suction pressure. The flow is affected by the suction difference and resistance to flow in the plant tissues, and it can be shown notationally that the quantity of water (W) flowing per unit of time is related to the suction difference and the inverse of the resistance (r):

$$W = \frac{Z}{r}.$$

The power involved in the flow of water in a plant system is given by the equation:

$$P = W \times Z.$$

In the plant world, many external factors influence this relationship, or it can be referred to as an open system in which plants play a part in *water management*. The principle of water management incorporates the fundamentals of suction, water flow and internal resistance. Once again, this can be shown to be consistent with other physical models for water management.

2.6. Animal Systems

The hydraulics described in Section 2.5 also hold good for blood circulation in animals, and full details can be read in an article in the *Transactions of the ASME*, September 1972, "A large-scale model of the human cardiovascular system and its application to ballistocardiography" (University of California).

3.

The Forms of Energy and Their Applications

"Thirty spokes radiate from the axle, but the emptiness in between forms the substance of the wheel" (Lao Tse).

3.1. Introduction

Instinctively, prehistoric man used three kinds of energy, namely heat, mechanical and chemical energy, without realising the connections between them.

1. *Mechanical energy* was used for manual work and people were a source of such energy for millions of years. Supplementation did not occur until 5000 years ago when the alternatives to muscle power were discovered. The very first source of mechanical energy was the muscles of men and animals; they were used to build the pyramids and to cultivate the fields in order to grow foodstuffs. Technical improvements were made by the discovery of the wheel for movement over the land and the oar and sail for movement across water.
2. *Thermal energy* was obtained directly from the rays of the Sun, or by friction when making fire. There is evidence that the people knew how to make fire 200,000 years ago; firstly in the home and then for technical purposes.
3. *Chemical energy* was available in natural forms as foodstuffs in plants, by photosynthesis (aided by chlorophyll and sunshine) and in animals.

These three forms of energy were used until recent times when they were supplemented by two other forms.

In the nineteenth century *electrical energy* was developed from chemical energy, initially, and later through electromagnetism in generators. At first, little was known about electrical energy, but playing with it awakened all kinds of new ideas that led to technical developments, such as the Morse telegraph in 1836 and the Edison electric lamp in 1880. Very soon electrical energy became a universal form of energy.

In the twentieth century atomic or nuclear energy resulted from the discovery of radioactivity. Einstein's formula $E = mc^2$ shows that mass at high speed transforms into energy, and it appeared that radioactive material lost mass, namely part of the atom itself, when it gave out energy.

Further studies led not only to the atomic bomb but to the possibility of using this source of energy for industrial purposes, using fissionable material. This is now an acceptable process that must grow and blossom in the future.

3.2. Applications for Energy

The first question to be asked is "What kind of energy is needed?" The second is "Which is available?"

In the Introduction it was mentioned that the original energy, which was used in prehistoric times, was fire and the conversion of chemical energy into thermal energy. Also, muscles that converted chemical energy into mechanical energy for doing work. Water and wind were other examples of mechanical energy that converted kinetic energy into work.

Other energy transformations were known in ancient times, including the conversion of friction (mechanical energy) into fire (thermal energy), as well as the reverse in the example of the "eolipyle". Hero of Alexandria designed this machine which was really a steam turbine that worked on the reaction principle. Although the "eolipyle" was well made, it was not viable at the time; later the turbine of Giovanni de Branco (1629), using a similar design, was little more than a sketch. So far, only technical considerations have been covered and not the practical use of energy. Transport over water, where only waves and currents had to become overcome with oars and sails, was easier than land transportation. Less brute force is needed, where only waves and currents have to be overcome. Early man must have realised this and sea voyages were recorded 1000 years ago between the Gulf of Bothnia and the Mediterranean Sea via the British Isles. Similarly, there were early voyages from Europe to America and across the Pacific Ocean by the Polynesians.

Travel by sea with sails was advanced when the sails could be inclined to the force of the wind in order to travel in any direction; also, by the introduction of the pointed hull for reducing the drag of the water. A larger sail area gave more driving force to ships.

In the Middle Ages, water- and windmills were developed in order to harness these natural sources of energy. In Holland and other countries, they contributed to industrialisation — for processing milk products, oil seeds, rice, spices and timber. Elsewhere, there were mountains with fast rivers that were harnessed for driving watermills, particularly in the Black Forest. In flat countries, the cornmill was a common sight and some still function today.

3.3. The New Era of Steam Power

Development of steam power took place in England, but before this, technical progress had been made all over the world, just as much in China as in the Middle East. Although England was the centre for this terrific technical advance, it soon spread to the north and the west of Europe and to America.

England was fortunate in having coal seams that reached the surface in places and during the time of the fourteenth century coal was being used for household fires; strong regulations for preventing air pollution followed. Presumably, through improved methods of stoking fires and the economic pressures of urbanisation, the use of coal increased in Britain. The surface coalmines were unable to supply all the needs and people started to dig in the mountains and the classic danger of all mine-workings soon appeared — water.

Originally, coal was extracted with manual labour and later with the help of horses; however, it was evident that a more powerful alternative had to be found. At the end of the seventeenth century, the solution appeared when Savoury invented the steam-operated pump. It was a primitive machine that worked with a vacuum chamber which was filled with steam under pressure. Then, the steam was condensed with cold water and the chamber sucked up water. This could be repeated continuously. It was simple and had no moving parts, yet it worked, although most inefficiently. In operations, heat and cold were trapped in a vacuum and it used coal which was both cheap and plentiful. And so the era of steam power began.

Soon there followed an important improvement — the machine of Newcomen which was also based upon a vacuum. The vacuum separated the steam from the water pumps and it worked with flap valves in its cylinders. After this, Watt suggested an independent condenser and later designed a steam engine to use higher pressure, incorporating a rocking beam that could turn a shaft. When he designed a regulator, it opened up the field of technical control.

The transformation of chemical energy into thermal energy in the boiler fire was followed by a machine that converted this thermal energy into mechanical energy, a very fruitful outcome. In the boiler, thermal energy passed from the fire into the water in order to heat it and generate steam for the pump. The water and steam operated in a closed circuit.

Steam from the boiler did work for pumping water and then it was condensed back to water and returned to the boiler. All later designs were improvements on the original principle. For example, steam pressure and temperature have been raised and the steam has been superheated, all of which extended man's technical knowledge.

Developments that followed included: casting and smithing techniques; metallurgy and a better knowledge of material properties; pressure sealing and lubrication. Superheating led to the introduction of mineral lubricants that could resist high temperatures.

In the second half of the nineteenth century there were two applications of steam turbines that have remained until today. The time was ripe for these steam turbines to open the way for other industrial developments, among them the piston engine. The ground was prepared now for the use of mechanical energy on a large scale and the Industrial Revolution followed. Textile machinery, steamships and railway locomotives changed the face of society with enormous consequences, such as rapid economic growth and expansion, social aggrandisement and casting off of the feudal system. The West eclipsed the rest of the world, and much use was made of it. These developments were possible through the coincidence of three factors:

1. An existing level of technology (technical qualifications).
2. The development of suitable solutions that left room for improvement.
3. The availability of mineral resources, particularly, coal and iron (thermal energy).

Coal is a derivative of wood, but it has double the thermal value; however, the time was right for its exploitation, and its use was instrumental in the growth of commerce.

3.4. After Coal: Gas and Oil

It is particularly worth mentioning that industrial development was fastest in North America where they had everything in their favour: fantastic mineral resources, freedom in society and social incentives without the shackles of tradition. In the middle of the nineteenth century Pennsylvania was a hive of activity, anything was needed and everything was tried. The climax was reached when Titus Drake found petroleum oil in 1859 when he was drilling for water. It was suitable for gas lighting and the demand for this fuel soon spread widely throughout the land as a welcome substitute for gas which was a product of coal. Also, another by-product of distilling crude oil was suitable for lubricating oils, especially in the high-temperature operating conditions of the steam engine.

Despite these new developments, the tempo of industry was still slow and there was no indication of the explosion that was to follow. Nevertheless, the signs presented themselves and quickly the success of the Industrial Revolution that was based upon the steam engine stimulated people in smaller industries to look for a simple solution to their problem of increasing output.

The steam boiler required daily attention and it was potentially dangerous; therefore, a search for its replacement started. It appeared in the form of the piston-operated internal combustion engine that burnt petroleum gas. The force of the explosion caused by this combustion pushed a piston down a cylinder in much the same way as the steam engine of Newcomen. The piston engine gained a new lease of life.

Actually, the breakthrough was made by a Frenchman, Beau de Rochas, who put forward the idea that the piston should operate on a four-stroke cycle: suction — compression — combustion — expansion.

It took a German, Nikolaus Otto, to make the idea into a practical success, especially when it was realised that any inflammable gas could be used, from volatile liquid fuels and methylated spirits to a product of the gas industry called benzole.

3.5. People on Wheels

Yet another desire awaited fulfilment. A machine was needed in order to free people from the hardships of doing manual work. Efforts to design a suitable small steam engine were unsuccessful, but the engine of Nikolaus Otto was a step in the right direction. Soon, the pioneers Daimler and Benz made their first horseless carriages.

There was great scope for improvement, and different companies led the field at different times. The uneven surfaces of roads were disastrous to every iron carriage of the nineteenth century until Dunlop in Ireland made a pneumatic tyre for his bicycle. Quickly, automobile manufacturers saw the light. The motor car lacked comfort due to vibrations of the road and uneven combustion in the engine, which was prone to producing clouds of oily smoke. Even so, its following was great at the turn of the century in Germany, England and the USA.

Now danger was being seen as adventure. From the manufacture of kerosene, a very volatile and easily evaporated by-product which was useless for burning as a normal fuel was found to be excellent for combustion in the Otto engine and, thus, gasoline and the engine found each other.

It only required some technical development for the motor car to be joined by the motor cycle and for production methods to be improved in order to comply with the demand. In the USA prominent men who contributed to this were Taylor and Gilbreth in the field of industrial engineering and Ford in the field of mass production. Henry Ford made a simple and robust vehicle from the best materials, although it became known as "The Tin Lizzie". It was the first motor car to reach the manufacturing figure of one million, and it opened up the countryside, firstly in America and then in many other countries after the First World War.

The effects on society were greater than anybody had dreamed, with diversions up side roads to new applications; however, all were based upon the old concept of converting energy in the sequence of chemical-thermal-mechanical energies.

The diesel engine appeared in 1893, after many years of development; it was almost perfect for agriculture, transport and contracting purposes, propulsion of ships and generating electricity in remote areas. Its fuel required less distillation from crude oil and it was less inflammable than gasoline.

3.6. People with Wings

By the turn of this century, Otto's engine had been mounted in an aeroplane, although it crashed, but the first powered flight was made a few years later by the Wright brothers. When sitting in a Boeing 747 one has to realise that this first aeroplane played a part in the manufacture of modern aircraft.

Shortly after the Second World War, the development of the aircraft engine culminated in versions with 18, 24 or 28 cylinders, because the maximum attainable power per cylinder had been reached. Provisions had been made to soar to great heights and in very thin atmospheres, but the volumetric efficiency of an engine prevented it from exceeding about 4000 hp. The four-engined aircraft was the most practical version and it helped to develop the world's airlines in the decade after the war.

At this culmination point, a successor appeared in record time — the jet engine, designed by Frank Whittle.

The breakthrough of the steam engine in the twentieth century could be attributed to the volumetric efficiency of the turbine to a certain extent and because the steam turbine was capable of delivering greater power. Likewise the turbine converted the internal combustion engine into a jet engine with a far greater power output. As with the steam engine, the positive effect was apparent from its earliest days, although its thermal efficiency was poor. Its higher demand for fuel is a disadvantage only when fuel is scarce, when more attention has to be paid to improved efficiency.

The vacuum principle used by Watt, whereby steam under high pressure expands, had an efficiency of only 5%, but after condensation returning a little of the water to the boiler improved the efficiency in some way or another. The gas turbine, on the other hand, compressed the air that it used.

In the first part of the century, rotary compressors were very inefficient, as opposed to the suction type of internal combustion engine or steam engine, where the only limitations were the properties of the gas itself. This restricted the permissible temperatures which the gas turbine overcame with a closed circuit.

In conclusion, it can be seen immediately that the gas turbine must have a much greater volume of air than it needs for combustion. The merit of

Whittle's design is that it utilises aeropropulsion, whilst the classical aircraft engine uses a propeller that displaces air without it playing a part in the propulsion. Whittle showed that a light compact engine performed no work other than the blasting out of gas at high pressure in order to produce a reaction force that propelled the aircraft through the air.

Against disbelief and opposition, he built such a small engine using very high quality materials and it proved itself to be a great success in the Second World War. In production, it received much praise and helped to combat the notorious V–1.

Germany also had operational jet fighters before 1945, and after the war both concentrated on the gas turbine so that it had a big impact on air traffic. Before 1955, the piston engine had been dethroned by the obvious advantages of the new engine:

— a lighter engine without a propeller,
— a great power output (around 20,000 hp),
— much greater speeds and larger aircraft,
— higher flying into the thinner atmosphere which has less friction,
— no vibration and less noise in the cabin,
— a better aerodynamic shape (Caravelle),
— an unforeseen simplification of maintenance.

Continued success was due to yet increased airflow for propulsion, by utilising a high-speed fan this time (the by-pass or fan engine). This had economic advantages and greater efficiency with a better take-off distance (shorter runways) and the opportunity to develop new materials like titanium and carbon fibres. The development of the gas turbine for other uses increased its success, but its operation still remained limited. For example:

— electricity generation is more efficient with a steam turbine; it gives more power and uses a cheaper fuel; but the gas turbine is suitable for peak loads, because it reaches its rated power output quicker than a steam turbine;
— automobiles — higher price and lower response-time than a piston engine;
— ships — starting with container ships because of smaller machine volume requiring less maintenance.

3.7. Atomic Energy

The fantastic success of different combustion engines increased the consumption of oil and natural gases and the spectre of their exhaustion soon loomed over society. One thing and another stimulated the development of nuclear energy, based upon splitting the uranium atom in order to produce heat. So here was the most recent development in the use of thermal energy.

3.8. Water Power

Quantity

The world reserves of hydrostatic energy amount to 3×10^{12}W, of which 10% is readily available (2.1×10^{11} in 1964); however, it is poorly distributed over the Earth's surface. The greatest reserves are found in Africa, and 20% is used in America, although this could rise to 50%. The reserves in Europe are relatively small and, since the most accessible supplies of water are being used already, any improvements will have to be quite sophisticated.

In many places larger installations are being constructed. For example, the Grand Coulee Dam in the Northwestern United States is capable of producing 9000 MW of electricity and will be the largest power station in the world.

3.9. Solar Energy

For small-scale applications, solar energy can be used in many countries, but research in this field is poorly co-ordinated. Recently, proposals have been made for great use of solar energy, but they had such contrary objectives that progress has been slow.

Equipment for remote areas in the developing countries tends to be in the "do-it-yourself" category, whilst other people have proposed huge satellite space stations for beaming solar energy to Earth. The choice between centralised and decentralised systems has caused still further conflict.

Solar energy problems include the uneven distribution of sunshine over the Earth's surface, the low density of energy falling on a unit area, the large collection surfaces, energy storage methods and suchlike. The big advantage of using the Sun's energy directly is its lack of disturbance on the balance of nature.

Quantity

The amount of solar energy falling on the Earth is 1.8×10^{14} kW, which is 10^4 times the present-day energy consumption. *The Sun is the only energy source which can provide the human race with a permanent supply of energy.* The Sun provides a global radiation source of 6000K; as a rule of thumb, in The Netherlands, 1 kW per m^2 falls perpendicularly on the soil and 100 W per m^2 horizontally (the maximum in summer is 200 W m^{-2} and the minimum in winter is 20 W m^{-2}). In fact, the amount of energy falling on 1.4% of the surface of The Netherlands was equal to the total energy consumed in 1970.

The amount of energy that a family requires falls on the roof of its house. When the Sun is overhead on a sunny day, the energy received is between 200 and 300 W m^{-2}. The maximum use of solar energy for household consumption is provided when the house faces southwards, or rotates with the Sun.

4.

Theoretical Rules for the Energy Concept

"There would be no space or time if there was not mass" (Einstein).

4.1. Physical Systems

The word "energy" derives from the Greek *energon* that translates into "source of activity". The philosopher Heraclitus (*c* 500 B.C.) has used the word when he pronounced upon *panta rhei*, that is "everything flows"; however, flow requires metamorphosis and favourable conditions. Much later, the word "energy" was introduced into the West by De Bernouilli (1725) who used it to describe the product of force and distance.

He associated it with a system of levers to show that a small force could lift proportionally larger weights, the proportion being the inverse ratio of the respective distances from the fulcrum. In the nineteenth century, equivalent forms of this energy concept were demonstrated, and credit for this is ascribed to J. R. Mayer (1814–78). Mayer calculated the mechanical equivalent of the calorie by the difference in specific heat of gases at constant pressure and volume. J. R. Joule introduced equivalents for heat and electricity into the scheme. By a process of induction, attempts were made to form a basic energy theory.

How was it possible that energy in one form could disappear only to reappear elsewhere as another form? There seemed to be no medium for energy transformations; however, learned men (including Maxwell) proposed a hypothesis that there was a medium that penetrated all matter, and they called it "the ether". Then, Einstein concluded that energy is independent. Physical scientists distinguished different forms of energy, like kinetic, potential, mechanical, acoustic, hydraulic, pneumatic, aerodynamic, torsion, resistance, osmotic, chemical, thermodynamic, electrical, electromagnetic, radiation, atomic, anti-material and others.

Regarding the formulation of a theoretical concept of energy, many ideas have been put forward, including the general theory of W. Heisenberg *et al*. His theory can be stated as: "In the universe energy is eternal; it is present in a constant amount although it can be changed in form. Energy has three fundamental characteristics:

1. *Persistence*, which is the tendency for energy to persist in a certain state . . . (p).
2. *Extension*, which is that part of the system that is the medium for its persistence . . . (e).
3. *Time* . . . (t).

From these three factors, others can be deduced:

4. *Action*, which is $\frac{de}{dt} = v$.
5. *Impulse*, which is $p.v = I$.
6. *Force*, which is $\frac{d}{dt}(p.v) = F$.
7. *Energy*, which is the product of force and extension: $F.e = E$.
8. *Power*, which is $\frac{dE}{dt} = \frac{\mathrm{d}}{\mathrm{dt}}\int F.\, de = \frac{d}{dt}\int F.v.dt = F.v = \mathrm{P}$.

Table 1 gives a framework for the proposed relationships, based upon the survey of well-known magnitudes.

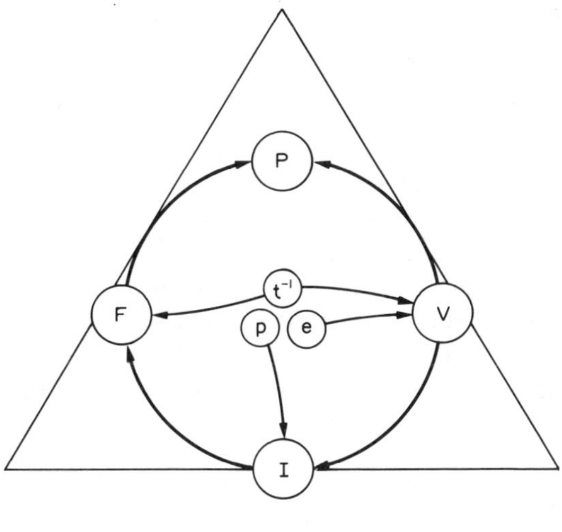

FIG. 2. Basic paradigm for power.

Figure 2 shows the connections between constituents and the units of action, impulse, force and power which will be used as guidelines in the following chapters. The triangle enclosing the circle symbolises that power always functions within a system.

Table 1.

Dimension System	Persistence $[p]$	Extension $[e]$	Time $[t]$	Action $[v]=[et^{-1}]$	Impulse $[pv]=[pet^{-1}]$	Force $[F]=[pvt^{-1}]=[pet^{-2}]$	Power $[P]=[Fv]=[pe^2t^{-3}]$	Energy $[E]=[PT]=[pe^2t^{-2}]$
Hydraulic	hydraulic inertance	volume	"	volume flow	hydraulic momentum	pressure	power	energy
Linear	mass	length	"	velocity	momentum	force	power	energy
Rotational	moment of inertia	angle	"	angular velocity	angular momentum	torque	power	energy
Light	Planck's constant	cycle	"	cycles per time	photon	force of radiation	radiation flux	radiation energy
Chemical	affinity	massflow	"	reaction speed	flux	chemical potential	power	energy
Electrical	self-inductance	electric charge	"	electric current	magnetic flux	voltage	power	energy
Magnetic	electrical flux	magnetic flux	"	magnetic flow	electric charge	magneto-motive force	power	energy

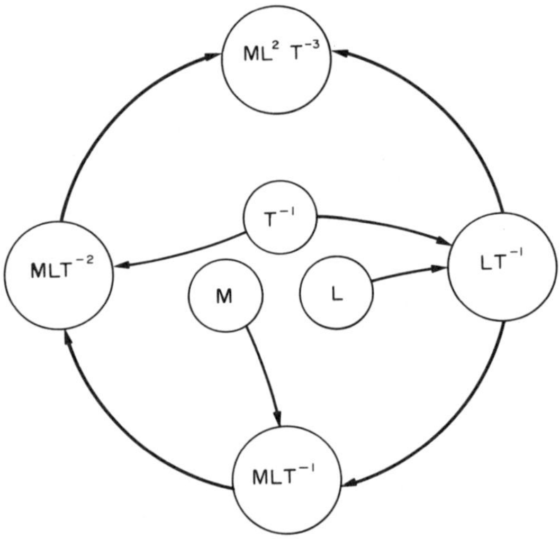

FIG. 3. Paradigm for power with *M, L* and *T*.

Figure 3 illustrates the general paradigm for power using the basic dimensions *M, L* and*T*.

The dimensions of a unit indicates its characteristic, whilst the mutual connections between units in the fixed system are expressed in terms of the dimensions given to the proposed basic characteristics. From classical mechanics, the basic unit of mass (*M*) is known; also, length (*L*) and time (*T*). Returning to the energy concept, these units can be used to express different derived units:

Velocity (*V*) $= LT^{-1}$
Impulse (*I*) $= MLT^{-1}$
Force (*F*) $= MLT^{-2}$
Power (*P*) $= ML^2T^{-3}$

Dimensions of derived units can be established such as for surface flow, hydraulics, torsion, volumetric elasticity and crystallography. They can all be expressed in terms of M, L and T.

In the *case of surface* flow, the surface area is the basic dimension unit. The basic dimensions are:

Areal mass	$= ML^{-2}$
Surface	$= L^2$
Time	$= T$

The derived dimensions are:

Surface velocity increase	$= L^2T^{-1}$
Surface impulse	$= MT^{-1}$
Surface tension	$= MT^{-2}$

In the *case of hydraulics*, the volume is L^3 and the basic dimension units are hydraulic inertance $= ML^{-4}$, volume $= L^3$ and time $= T$. Then, the derived units are:

Volume velocity	$= L^3T^{-1}$
Hydraulic impulse	$= ML^{-1}T^{-1}$
Pressure	$= ML^{-1}T^{-2}$

Torsion in thin bodies refers to the four-dimensional changes and the polar quadratic surface moment (L^4) is used. It is the product of the resistance moment (L^3) and the greatest distance (L) between the axis and a point on the cross-sectional area.

Volumetric elasticity refers to bodies that deform when pressure is exerted on their outer surfaces. Each deformed body reacts to a pressure that is proportional to its surface area; therefore, the basic dimension unit is L^5 which is a product of volume (L^3) and the cross-sectional area (L^2).

Crystallography is concerned with the growth of crystals. Growth starts with a germ and proceeds three-dimensionally, i.e. a three-dimensional increase in volume, or a product of L^3 and L^3.

Table 2 shows the relationship and dimensions of the units.

Table 2.

System	Basic components			Derived components			
Crystal physics	T	ML^{-10}	L^6	L^6T^{-1}	$ML^{-4}T^{-1}$	$ML^{-4}T^{-2}$	ML^2T^{-3}
Volume elasticity	T	ML^{-8}	L^5	L^5T^{-1}	$ML^{-3}T^{-1}$	$ML^{-3}T^{-2}$	ML^2T^{-3}
Torsion	T	ML^{-6}	L^4	L^4T^{-1}	$ML^{-2}T^{-1}$	$ML^{-2}T^{-2}$	ML^2T^{-3}
Hydraulics	T	ML^{-4}	L^3	L^3T^{-1}	$ML^{-1}\mathrm{T}^{-1}$	$ML^{-1}T^{-2}$	ML^2T^{-3}
Surface of liquids	T	ML^{-2}	L^2	L^2T^{-1}	MT^{-1}	MT^{-2}	ML^2T^{-3}
Classical mechanics	T	M	L^1	L^1T^{-1}	MLT^{-1}	MLT^{-2}	ML^2T^{-3}

The hierarchy increases dimensionally from L^1 to L^6 and the sequence of these energy forms can be formulated thus.

$$E_n = (ML^{2-n}T^{-2}).(L^n)$$

This equation can be expressed simply as the product of a force ($ML^{2-n}T^{-2}$) and an extension (L^n). Subsequently, the dimensions of other extensions can be described with the three spatial dimensions, L^1, L^2 and L^3, where L^1 is the length dimension, L^2 the area dimension and L^3 the volume dimension. Actually, the progression $L^1 \ldots L^n$ is a repetition of these three spatial groups.

1. The first group comprises:
   — translation with extension (L^1),
   — surface flow with extension (L^2),
   — hydraulic extension (L^3).
2. The second group comprises:
   — torsion ($L^1.L^3$),
   — volumetric elasticity ($L^2.L^3$),
   — crystallography (L^3).(L^3).

Figure 4 will serve as a visual aid for understanding these assumptions.

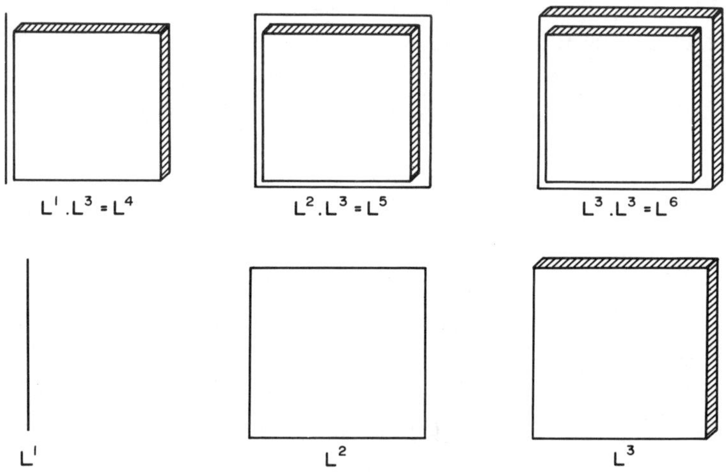

FIG. 4. Spatial dimensions.

By rotating a body through an arc, we arrive at a dimensionless extension, the angle. Angular velocity for uniform rotary motion is the quotient of the angle and the elapsed time for the motion and it is measured in radians per second.

The dimensionless radian can be described as the extension $L°$; therefore, angular velocity becomes $L°.t^{-1}$. The moment of inertia of a rotating body refers to the amount of resistance to rotary motion about an axis, e.g. the persistence of rotary motion. The unit is represented by kg.m^2 where the notational dimensions are ML^2. Simply speaking, a small mass (m) rotates through an arc that can be described as the difference between the perpendicular normals of two tangents (see Figure 5).

In the case of rotary motion, the following units are involved:

Time	$= T$
Persistence	$= ML^2$
Extension	$= L^\circ$
Action	$= L^\circ.T^{-1}$
Impulse	$= ML^2T^{-1}$
Force	$= ML^2T^{-2}$

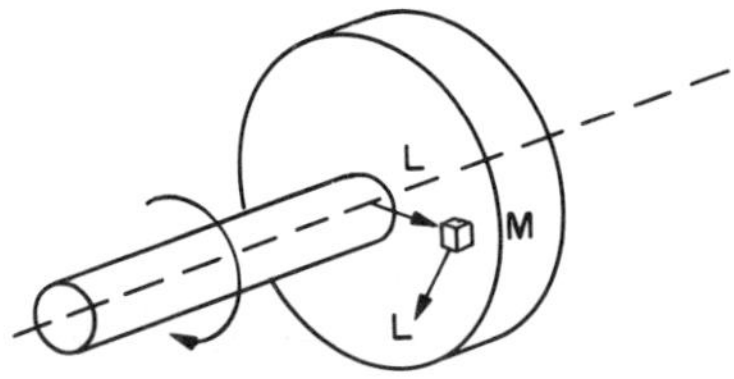

FIG. 5. Rotary motion.

The dimensions of persistence for electrical, nuclear and chromodynamic phenomena give rise to expressions for two (rotation and spin of electrons), three (spin of nuclei) and four (spin of quarks of protons) rotations that are fundamental to these phenomena. Using the Theory of Relativity propounded by Einstein, the bending of light beams can be seen as a counterpart of tension; therefore, it can be given a dimensional unit of L^{-4}.

Pulsation of the universe may be the counterpart of volumetric elasticity and its extension unit can be written as L^{-5}. The concept of a universe being created from the explosion of a primeval atom relates back to an extension unit of L^{-6}, which could be seen as the counterpart to the extension for crystallisation.

The following dimensions for the basic and derived units (Table 3) could be allocated to the above-mentioned forms of energy.

Table 3.

	Time	Persistence	Extension	Action	Impulse	Force
Electrical energy (rotation and spin of electrons)	T	ML^4	L^{-1}	$L^{-1}T^{-1}$	ML^3T^{-1}	ML^3T^{-2}
Nuclear energy (spin of atom-nuclei)	T	ML^6	L^{-2}	$L^{-2}T^{-1}$	ML^4T^{-1}	ML^4T^{-2}
Chromodynamic energy (spin of quarks of protons)	T	ML^8	L^{-3}	$L^{-3}T^{-1}$	ML^5T^{-1}	ML^5T^{-2}
Bending of light beams	T	ML^{10}	L^{-4}	$L^{-4}T^{-1}$	ML^6T^{-1}	ML^6T^{-2}
Pulsing of the universe	T	ML^{12}	L^{-5}	$L^{-5}T^{-1}$	ML^7T^{-1}	ML^7T^{-2}
Primeval atom	T	ML^{14}	L^{-6}	$L^{-6}T^{-1}$	ML^8T^{-1}	ML^8T^{-2}

4.2 Biological Systems

In Chapter 2 it was mentioned that ATP is one of the most universal elements of life; also, that energy forms the matrix of life. Obviously, all the species of plants and animals are closely related to energy.

From the crystal form of energy, the next step is to the living forms, starting with the virus. Viruses cause infections in man, animals and plants. The unicellular viruses are too small to be seen under an ordinary microscope and they can be distinguished only with an electron microscope. They have different shapes and, chemically, they consist of DNA or RNA covered with a layer of protein.

A virus can reproduce itself only within a cell of its host, outside of which it does not grow or show any form of activity. Under favourable conditions, one virus can divide into two or three thousand new specimens.

Plants and living creatures are composed of *cells* that are so small that they cannot be identified without a microscope. Each cell of a plant has an outer wall that consists of cellulose, and it encloses the protoplasm which is the living part of the cell; from it the cell wall itself was developed.

The protoplasm is a slightly sticky fluid in which can be seen a number of small bodies, the largest being the nucleus, which acts as the co-ordinator of the living process in the plasma; removing it will kill the cell in a short time.

In higher organisms, the living process within cells relies upon DNA, and it has been isolated from the cells of many different species of plants and animals. However, microchemical research, with the aid of radiographs, shows a great similarity of the composition and structure of DNA molecules from different species. These molecules appear to comprise double chains of linearly joined units called nucleotides (see Fig. 6). Each unit is made up from phosphoric acid, a sugar (desoxyribose), and has a nitrogen base. There are four different nucleotides and the only means of differentiating them is through chemical analysis of the base substance. Normally, the base is a purine (adenine or guanine) or a pyrimidine (cystosine or thymine). Individually, the base substances have been identified and given the letters A, G, C and T.

In composition, RNA is very similar to DNA, containing the same sugar molecules but not desoxyribose which is the "D" of DNA but ribose (the "R" of RNA), whilst the nitrogen base thymine is represented by uracil (see Fig. 6).

The modern presentation of a DNA molecule is based upon the model constructed in 1953 by Watson and Crick, in which there were two chains of nucleotides joined together with hydrogen bonds to the nitrogen base. Always, a molecule A of one chain was linked with a molecule T of the other chain, whilst G was always linked with C. Furthermore, the two nucleotides were wound into a spiral so that the chains looked like a spiral staircase. This chemical staircase consisted of alternate groups of phosphoric acid and sugar and they displayed a paired formation of

(A–T) + (T–A) and (C–G) + (G–C) linkages (see Fig. 7).

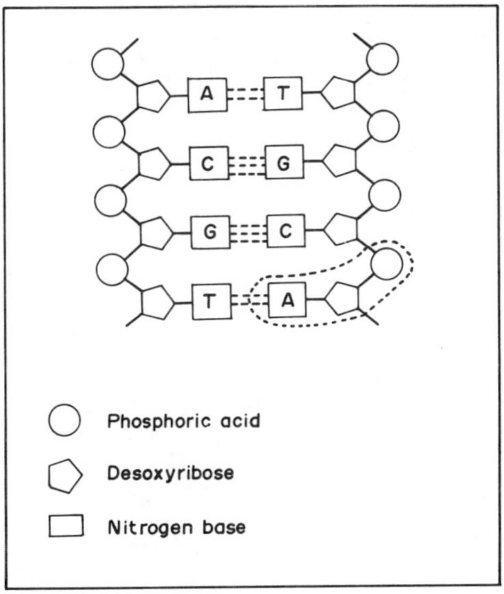

FIG. 6. Nucleotides.

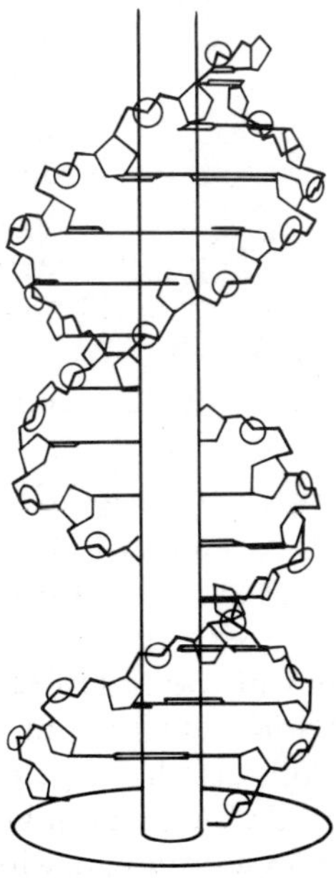

FIG. 7. Spirals of nucleotides.

Probably, the formation proceeded as follows: at the interface, when the nucleus divided, the double spiral of DNA unrolled. Then, an enzyme broke the relatively weak hydrogen bonds, so that the two chains moved apart from each other. Consequently, each chain of nucleotides came into contact with the surrounding nucleus fluid which contained the binding substance (a mixture of phosphate, sugar and the bases). A nucleotide base on one chain, then attracted a complimentary nucleotide base on the other chain (A ‹› T, T ‹› A, G ‹› C and C ‹› G). The new pairs became mutually linked and a new double chain was formed out of the original and they were identical. The sequence of nucleotide bases on each chain was the same; when the first was ACCAT, then the other one was TGGTA. The formation of chains is shown schematically in Fig. 8.

The word ACCAT gives the impression of being part of a code and, indeed, that is so. The code is built up from four signals, rather like Morse code which uses two different signals, a dash and a dot. When these two signals are combined, people can communicate with each other, and so it is with DNA. Probably, all hereditary factors are determined by the sequence of the four chemical signals, A, G, C and T. In the alphabet of RNA, the nucleotide base T is replaced by U. The combination of many thousands of bases makes it possible to produce a virtually endless number of DNA molecules, and consequently an endless number of genes.

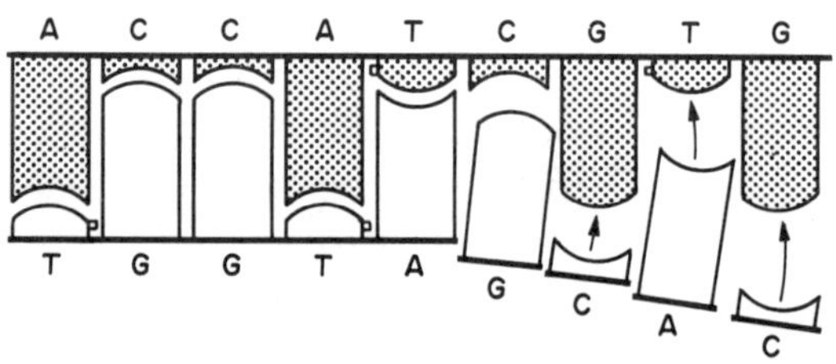

FIG.8. Formation of DNA.

As far as it is known, these DNA ladders are found in all living creatures, whether they are bacteria, trees, fishes, or human beings. Only the numbers and the frequencies of the rungs of the ladders are different; in other words, all living organisms display their own genetic information with the same alphabet of letters. They comprise the genetic code, and it has recently been decoded. Firstly, it was concluded that, for the synthesis of proteins, the genetic information was permanent and it could be "read" as words of three letters, e.g. AAA, CAG or UUU. Quite likely, a triple succession of amino-acid bases was coded, over and over again.

In this research, use was made of bacterial extracts in which the DNA chains were broken. These extracts contained ribosomes, enzymes, ATP (an energy source) and amino-acids — all the components for the synthesis of protein without the DNA and RNA which provided the instructions. In the first experiments, simple synthetic RNA preparations were added to the extracts which contained only U, A or C bases; consequently, the codewords were UUU, AAA and CCC. The original RNA contained only the amino-acid phenylalanine in a protein molecule which had the codeword UUU; the codewords for lysine and proline were AAA and CCC, respectively.

To date, the codewords for all twenty amino-acids are known, i.e. their letter combinations, although not the sequences for all of them. It follows that the gene can give instructions for the formation of amino-acids in the proteins. Also, it gives some idea of how the genetic characters are transferred to ribosomes in the cytoplasm which is where the proteins are "manufactured". The instructions originate from the RNA molecules and are carried by so-called RNA messengers, or m-RNA (see Fig. 9).

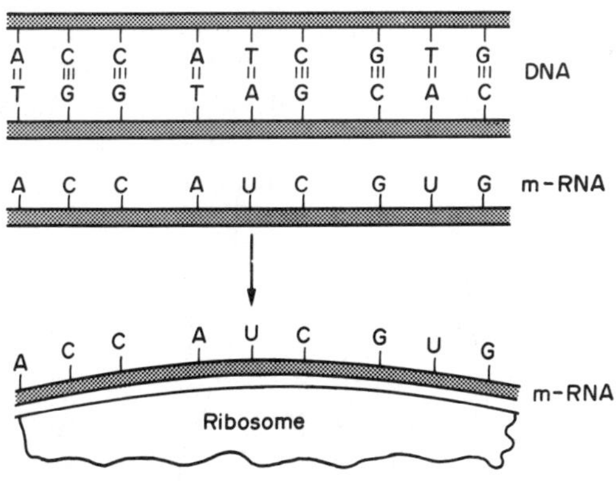

FIG. 9. Function of RNA-messenger.

When there is the need for a particular enzyme to be created, the gene which is responsible for its synthesis is activated. It makes an RNA molecule in its own image, but carrying the nucleotide base sequence that gives the instructions for building up the enzyme in question. Then, it leaves the nucleus and attaches itself to the ribosome molecule of the cytoplasm. However, it is not yet known which strand of the DNA chain forms the m-RNA.

Next, another type of RNA plays its part, the so-called transfer RNA or t-RNA. It transports the necessary amino-acid molecules to the ribosomes from the cell solution. Each type of amino-acid has its own t-RNA counterpart which it attracts chemically before depositing it in the correct place, with the aid of its three-lettered codeword, on the RNA molecule. When the RNA molecule is saturated, all its amino-acid molecules are then present in the correct order and the protein molecule is completely united. It is then released from the m-RNA and starts its function as an enzyme.

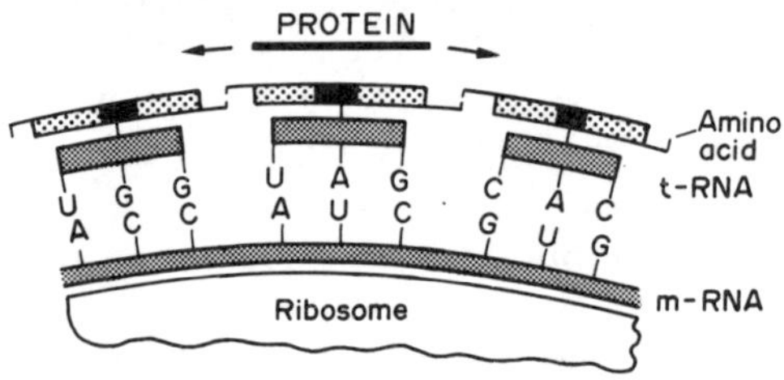

FIG. 10. Function of transfer-RNA.

After the t-RNA has separated from the m-RNA, the process can be repeated (see Fig. 10). Unless the life functions of all cells are co-ordinated, the organism cannot function as an individual. Indeed, the contents of a cell are not completely separated from its neighbours, because the cell walls have small apertures and the protoplasm passes through them in thin strands. The system of co-ordinated protoplasm is known as the symplasm. A collection of similar cells is called a *tissue*, and a collection of cells, like those that have just been described, is called parenchymatic tissue.

4.3 Differentiation

Many bacteria and algae are single-cellular *organisms* which are completely surrounded by their environment. In most instances, the environmental medium is water and the organism can take from it the chemicals that it needs for life.

Simple multi-cellular organisms have a high proportion of their cell walls in contact with the environment, but increasing the number of cells separates many of them from the outer layers. As an illustration, the surface area of a spherical organism increases as the square of its diameter, yet the contents increase as the cube of the diameter, as shown in Table 4.

When a cube of side 100 cm is composed of cells with side 1 cm, then a cell in the centre will be separated from the outside world by at least 49 other cells. All the materials that this cell needs for growth and function must pass through the cell walls of the intervening cells in the protoplasm. Now, the disadvantages of a cubic or spherical organism can be seen, and it has to adopt a method for increasing the surface area. The process of surface enlargement is common in nature, e.g. leaves of plants, but it is not the only solution. It is not suitable when greater strength is required, as with many advanced organisms. A leaf has a large surface area in contact with the air, but it needs water and mineral salts from the soil too. In order to transport them from the roots to the leaves, a great many cells would be involved, if the plant did not differentiate some of the cells into long tubes. They grow into a network to include all parts of the plants, so that a number of cells that were parenchymatic originally become differentiated for the purpose of transportation.

Table 4. Surface Area and Contents of a Cube and Their Ratios with Sides of Different Lengths (d)

d cm	$S = 6d^2$ cm^2	$C = d^3$ cm^3	$\frac{S}{C}$
1	6	1	6.000
2	24	8	3.000
3	54	27	2.000
4	96	64	1.500
5	150	125	1.200
6	216	216	1.000
7	294	343	0.857
8	384	512	0.750
9	486	729	0.667
10	600	1,000	0.600
20	2,400	8,000	0.300
50	15,000	125,000	0.120
100	60,000	1,000,000	0.060

As a result, the plant becomes a more complex organism and further parenchymatic differentiation will be necessary, such as strengthening the stems by laying down extra layers of cellulose to reinforce the cell walls and to make them woody.

Surface enlargement in land plants is accompanied by the risk of drying out easily and damage by high winds, for example. These risks are reduced by differentiating the outermost cells into an impermeable layer with a cuticle. The cells interlock like a jigsaw puzzle so that they possess greater strength. The process of cell differentiation for specialisation is very common in nature.

Numerous examples can be seen in the human body; the skeleton is made of very strong tissue called bone; blood vessels and nerves are transportation channels; the lungs display surface enlargement; and the skin or hide of animals provides a covering for the body in order to prevent drying out. After the crystal stage of evolution, there are successive developments of the unicellular life forms, of parenchymatic tissue, of flat structures like leaves and seaweeds, up through the higher lifeforms to man himself. The direction of all life processes in the cell is the responsibility of DNA, and its functions are controlled by nucleotids on the DNA spirals; they indicate a character of a persistence. Whereas it is the atomic arrangement that determines the configuration of a crystal, in DNA it is a continuous arrangement of a sugar, an acid and a base (in the nucleotides). This assumes that DNA has an extensional dimension greater than the crystal. When cells are arranged into a tissue, this assumes yet another dimension and a spatial organism yet one more. Therefore:

A *cell* extension = L^7; a *tissue* extension = L^8; an *organism* extension is L^9.

A further step in this evolution can be the development of nervous systems in which the three stages of measurement also can be identified – length, surface area and spatial volume. The nervous system is a quick-acting control system that provides information for the organism. Firstly, its sensory organs receive information from the outside world; then, they are integrated so that the functions of the various organs can be regulated and co-ordinated. Information is sent along the nerves in the form of coded impulses, where the number per unit of time is a measure of the capacity of the nerves. The speed of communication will depend upon the rate of impulses and the thickness of the nerve tissue; generally, speeds vary between 0.5 and 120 metres per second.

The linear phase is indicated by the differentiation of an epithelium cell into a neuron. A neuron is a nerve unit that receives and directs impulses. Directing is done with the aid of dendrons.

The surface area phase is demonstrated by the vibrissal family of worms. They have a network of nerve fibres and brain ganglia at their nodes.

The spatial volume phase is found in the vertebrates which have a spatial nervous system. The brain is its focal point and it controls all the functions of the body. With the development of the neuron, an organism obtains one more dimension; the ganglion system is another, and the brain is yet another. Therefore:

A *neuron* = extension L^{10}; a *ganglion* = L^{11}; the *brain* = L^{12}.

Sociopsychological Systems

In section 4.2 it was shown how phenomena of differentiation occurred in biology. Incidentally, they have their counterparts in sociology. In fact, throughout evolution repetitions of the same theme can be seen at different levels and it is easy to visualise the same pattern being repeated in sociopsychological systems. Within the biological system, the first group comprises the cell, a tissue and a complete organism; the equivalent in the social context could be the man, groups and a complete organisation.

The next successive group includes a business or industry, the branches of industry, and the national economic system (see the survey on page 38). In this organisational structure for energy forms, based on mass, space and time, persistence differentiates into ever-increasing complexity. This can be interpreted as an increased "immaterialisation" of persistence. An example is presented on page 36/37.

The dimensions of a sociopsychological system that may be synonymous with the proposed general energy system will be confined to four, about which additional support will be found in the literature.

1. Man — as described in terms of industrial psychology.
2. Group — as suggested by studies in group dynamics.
3. Industry — as seen by control theory.
4. National economic systems — as part of the macro-economy.

Literature about organisation and the branches of industry provides insufficient reference to points of contact for this grouping to be extended. On page 38, a diagram shows the groups of this system.

Structure

It is presumed that sufficient is known about the structures of the disciplines under consideration for a summary to suffice here.

Industrial psychology. The working functions of people in the production process have been described by Van Grunsven in *Bedrijfspsychologie* and A. K. Korman in *Industry and organizational psychology*, etc.

Group dynamics. The complex of lasting social relationships has been studied by Lammers en Van Doorn, *Moderne Sociologie, systemen en processen*, Robert K. Merton, *Social theory and social structure*, etc.

Industrial engineering. A structure with its geographic divisions and project or customer orientations for processes and products has been prepared by Prof. C. Botter, *Industrie en organisatie, bedrijfstypologieën*, H. A. Simon, *Administrative behaviour*, etc.

Macro-economy. Structures showing private and public investments and their diverse hybrids, structures for various systems of taxation, price-fixing and/or social control (J. Tinbergen, *Selected papers*).

Power

Physics is concerned with studying "power", mostly in closed systems, contrary to the social sciences which deal with open systems. Social sciences conform to the doctrine of *ceteris paribus* and within the range of certain thresholds functional relationships exist. The Cobb-Douglas production function can provide an insight into the macro-economic power model, whilst in the field of organisation theory industrial economics and group dynamics, important contributions were made by Ansoff, Van Grunsven, A. K. Korman, Lammers, J. G. March and G. Nielen. Also, the following articles provide further reading on this subject: Sherman, Roy and Krupp, "Structure of economics"; Buckley, "Modern systems research for the behavioural scientist"; I. G. and M. E. Wilson, "Information, computers and systems design."

The development of a physical model for power in the social sciences is assisted, on the one hand, by the work of Leibnitz, who reasoned that the whole of creation resulted from the transformation of energy and, on the other hand, by the systems discipline that incorporates many scientific principles.

An isomorphic view of the world suggests an objective direction to take, each aspect of which is a differentiation of concrete reality. However, we will confine ourselves to the power model of Cobb-Douglas, namely:

$$\text{GNP} = \text{capital} \times \text{work}$$

This model was expanded by workers in the school of Solov, Mead, Murray Brown and others — see their last model in which the influence of technical change on the production output is expressed as:

$$A = \gamma^{-1/\nu} \cdot B^{+1/\nu} \left[K \left(\frac{1-k}{k} \right)^{1-\sigma} \cdot p^{1-\sigma} + 1 - k \right]^{\cdot \frac{\sigma}{1-\nu}}$$

where A = production output, k = capital intensity,
B = GNP, ν = measure of increase,
K = capital inventory, p = related production cost factor,
γ = general efficiency, σ = substitution flexibility.

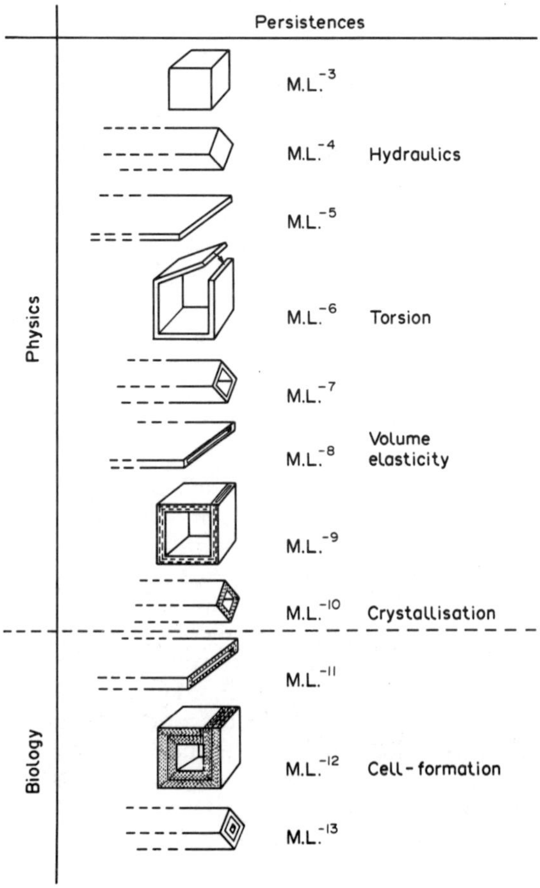

FIG. 11.

The explanation of macro-economic aspects should of course be completed with the trade cycle theory of Keynes and financial studies such as discounting and open market policies.

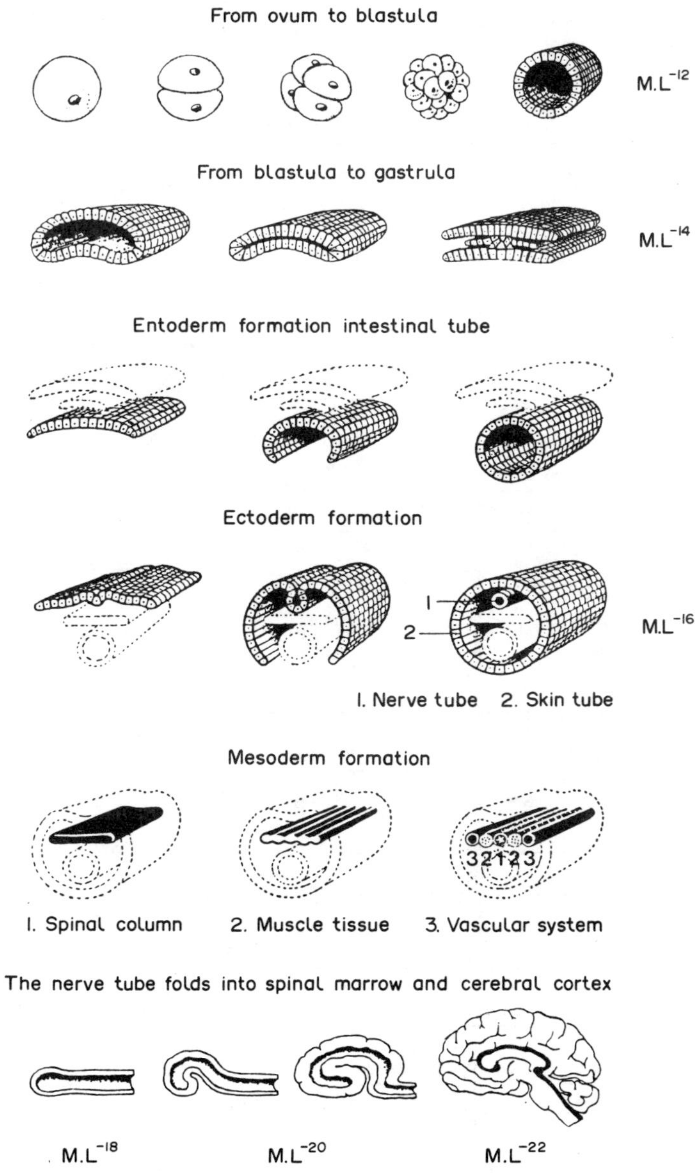

FIG. 12

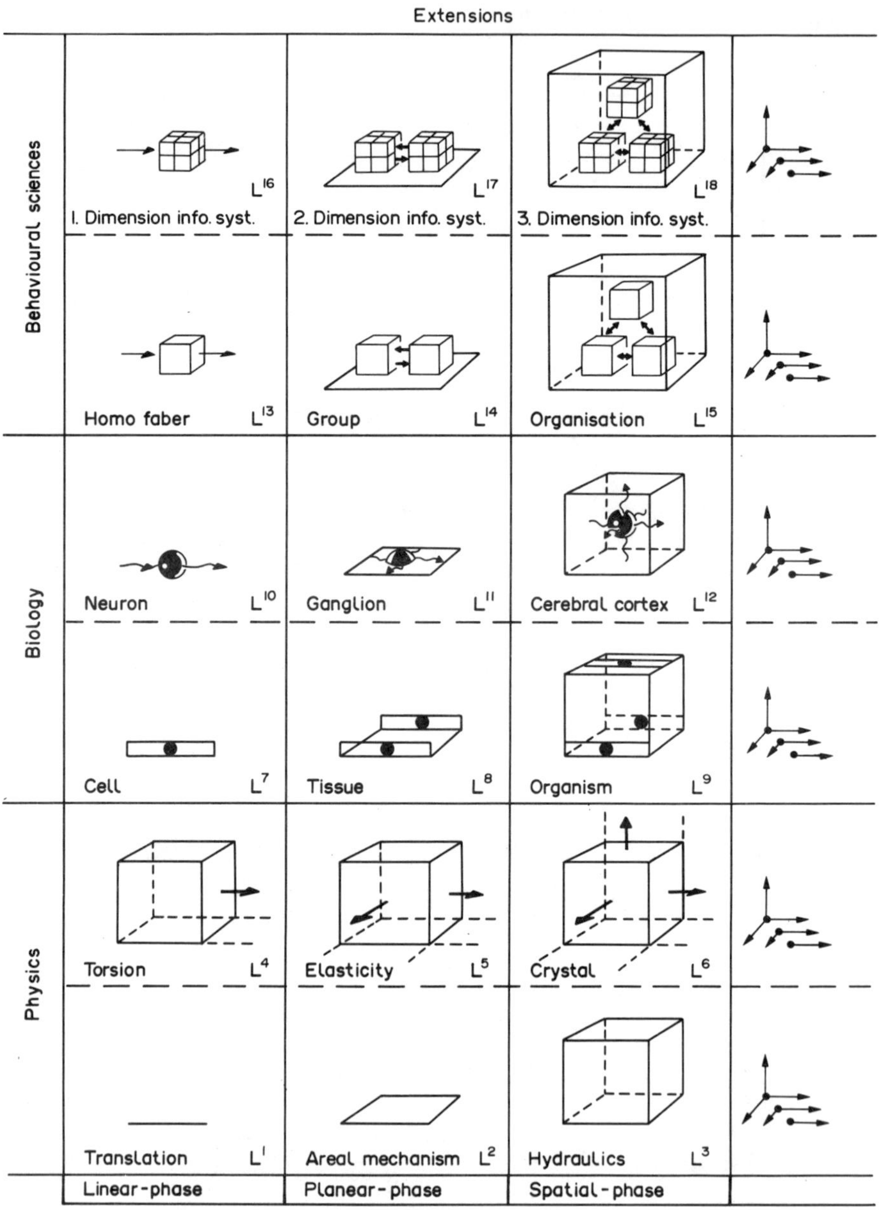
Extensions
Behavioural sciences
Biology
Physics
L16
1. Dimension info. syst.
L17
2. Dimension info. syst.
L18
3. Dimension info. syst.
Homo faber
L13
Group
L14
Organisation
L15
Neuron
L10
Ganglion
L11
Cerebral cortex
L12
Cell
L7
Tissue
L8
Organism
L9
Torsion
L4
Elasticity
L5
Crystal
L6
Translation
L1
Areal mechanism
L2
Hydraulics
L3
Linear-phase
Planear-phase
Spatial-phase

FIG. 13

5.

The Isomorphic Energy Paradigm for Industrial Psychology

"Each of us has a centre for initiative force and creative power" (Kant).

5.1. Introduction

All human work activities call for mental and physical efforts, usually directed towards production of goods or services. Industrial psychology has been carrying out research for a long time into the circumstances which influence the effectiveness of work, and a large part of it has been concerned with the differences between individual performances. In the following paragraphs we will look at the relationships between the performance of work, capacity, motivation and inherent capabilities in the isomorphic energy paradigm.

5.2. Performance (Power)

Vroom dealt with the notion of performance in his book "Work and Motivation" and he showed that it was possible to use several methods for measuring work in different situations. He suggested a scale for various levels of performance (power) showing the influences of ability and motivation.

5.3. Ability

The notions of ability and motivation include many considerations that affect attempts to measure them. However, when they are examined with the same terms of reference, it is possible to draw conclusions; for example, Vroom explained that "ability" was distinct from "performance". Considering a task in which the worker has been asked to respond to two signals, S_1 and S_2 only one response (R) will be the correct one. By definition, performance will be improved when the worker reacts more frequently to S_1 with R_1 and to S_2 with R_2. Experimentally, a scale for ability can be developed.

5.4. Motivation

In order to measure ability, work has to be done and for that it is necessary to motivate the worker. Research work has enlarged this observation, especially, the Hawthorne experiments at the Western Electric Company under the leadership of McGregor *et al.*

Motivation is defined by Buckley in *Modern Systems Research for the Behavioural Scientist*. He says that "motive is a trigger for the deed". What is a motive? It will be defined here as "*a hypothetical idea that evokes in a person conduct that has a predetermined direction and intensity*."

Some observations on this definition may include the concept of motivation as a hypothetical postulation rather than as something perceived. The objective is to decide the *modus operandi* for a particular purpose. Secondly, a motive calls for the determination of a plan of action. For instance, a person may feel a need to eat (the hunger motive) and he may think about cooking some food. Then follows a plan of action which includes looking into the refrigerator, removing some food, cooking it and eating it.

A third observation is that the stronger motives will stimulate the stronger actions.

A theory which assumes that personal motivation has a great influence on the psychology of work and organisation was proposed by Maslow, in which he classified the motives. He said that people continued in a motivated state (extension) although the grounds for the motive had changed; above all, people were never in a condition to appease their motives for very long. For example, the satisfaction of one need involves and develops other needs and Maslow suggested a hierarchy of needs in order to explain this better.

1. *Physical needs* form the strongest category. The satisfaction of this need takes highest priority for action. When it is the hunger motive, the need for food is not surplanted by the needs for particular kinds of food. In poor countries, the absence of food motivates people to eat anything; similarly, the safety motive often predominates over other needs.
2. *Safety needs* are associated with protection from danger or risk and from the threat or challenge of fear. In our culture, it is not so much a motive for protection as the resistance to physical harm. In an industrialised society, safety is a major concern for the relations between employers and employees, particularly conduct that provides no insurance against discharge, redundancy or bankruptcy.
3. *Social needs* are important only after the physical and safety needs have been satisfied. Social needs include the desires for friendship, the respect of others.
4. *Ego needs* include the need to achieve, the need for self-reliance and the needs for personal status and recognition.

5. *Self-actualisation needs* refer to the realisation of personal ability, which depends upon a continuous process of self-development.

Maslow suggested that the different motives overlap one another so that they were inter-related. A higher motive in the hierarchy arises before a lower one is completely satisfied. These relationships are presented in Fig. 14.

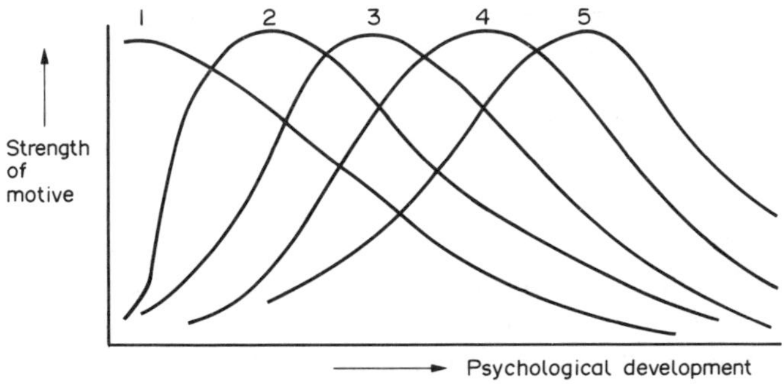

FIG. 14. Hierarchy of needs according to Maslow.

The Model of Vroom

Vroom performed investigations in the field of work performance, ability and work motivation. He showed that the effect of motivation on the level of performance depended upon the ability of the worker and, likewise, the effect of ability on the performance of work depended upon the motivation of the worker.

$$\text{Performance} = \text{Ability} \times \text{Motivation} \quad (1)$$

5.5. Perception

Introduction

Psychologists are interested in the way that people perceive the world about them and the way that they obtain information about it, via their sensory organs, and analyse it so that it becomes knowledge. Conduct only follows after sensory perception and it is a starting point for studying behaviour.

The sensory organs are concerned with perceiving distance (sight and hearing); influence on the skin (touch, taste, temperature and pain); and depth (position and movement of muscles and joints, sense of balance and awareness of internal organs).

The psychological process is the following:

Detection — Discrimination — Interpretation

Detection

Before a stimulus can play a part in human conduct it has to be perceived, or detected. In a dark room, a light point must have a certain intensity before it can be discerned; similarly, sound has to be loud enough to be heard. Consequently, the stimulus must have a "threshold value" before it can be detected.

Detection must account for the levels of stimulus that can, or cannot, be perceived and its measurement is part of the study of psychology. Psychologists investigate the sensitivity of the sensory organs and the effect of stimuli once they have been detected. It has been concluded that energy is needed for detection and its measurement is important for recording perception.

A number of tests have been devised for determining the "threshold values" for the different kinds of stimulus and for producing scales. When the stimulus energy is reduced, detection declines and it is possible to estimate the threshold value statistically.

In Fig. 15, the graph depicts the physical energy along the x-axis and along the y-axis the percentage of the persons tested who perceived the stimulus. It appears that the chance of perception increases with the amount of stimulus energy according to a particular function. This function is called *the psychophysical function*, because it expresses the relationship between a "psychological" variable (perceiving a stimulus) and a "physical" variable (intensity of the stimulus).

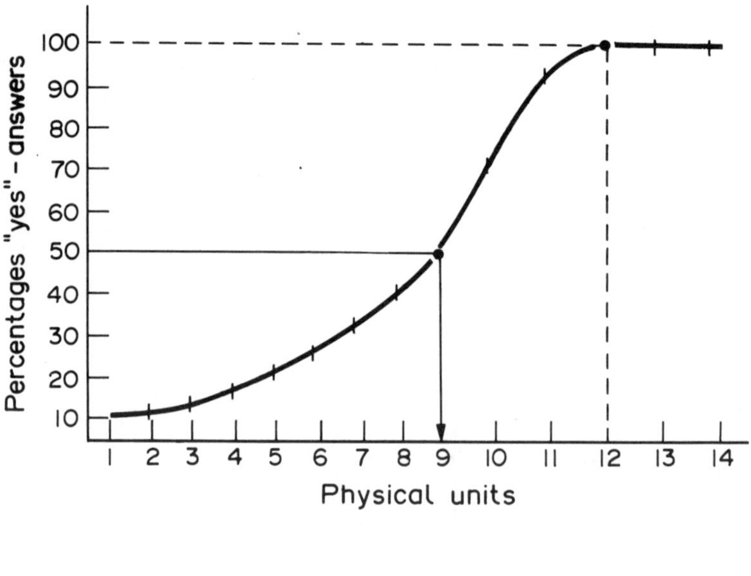

FIG. 15. The psychophysical function.

When the stimulus was less than three units, its presence could not be detected, and when it was above twelve units it could be detected always. Between three and twelve units the relative frequency of detection increases progressively. It has been accepted to define the threshold value as the value of a stimulus that can be detected by 50% of the people tested. In this example, it was given a value of nine. A number of threshold values are presented below. One has to realise, however, that the values vary from person to person and from time to time according to circumstances.

Some Threshold Values

Sight:	A burning candle at a distance of 50 km on a clear winter night.
Sound:	A ticking clock in quiet surroundings at a distance of 6 m.
Taste:	A teaspoonful of sugar in 2 litres of water.
Smell:	A drop of perfume distributed over a small room.
Touch:	The wing of a fly falling on the cheek from a height of 1 cm.

Theory of Signals Detection

The earlier experiments gave no explanation for variations in threshold values, but in 1954 Swets and Tanner applied the Statistical Decision Theory in order to develop a theory for it. They proposed, in general, that signals must be detected above a background noise, a background noise caused by an external source (away from the people) and an internal source (by the people themselves). Further information on these noise sources can be found by referring to Welford (1968).

Similarly, sonar-detecting instruments can tell the presence of a submarine from other irrelevant signals. With reference to the Theory of Signals Detection, detecting a signal has two essential aspects: (a) sensitivity and (b) deciding.

An example. Employee A reported a "strange" noise from a machine more often than employee B. Now the question is whether he is indeed more sensitive than B. It is of interest to note that: (1) *Learning* plays a part in hearing sounds which may have been imperceptible earlier. (2) Motivation and fatigue factors influence what people can hear.

Differentiation

Included with discrimination (differentiation) is the process by which one stimulus differentiates from another; it implies identification of the stimulus. Now there are two important problems.

1. *There must be a minimal discernible difference between two stimuli.*
   This is called the *threshold difference*, and two kinds of red colour must show distinctly different wavelengths in order to make it possible

to distinguish them apart, or two notes will have to record different decibel levels. Thus, for differentiation are necessary:

— *absolute thresholds* for detecting the presence of a signal or not;
— *difference thresholds* for detecting a difference between two signals.

It is interesting that the ratio of the difference threshold to the stimulus intensity is always constant. If the difference threshold for a 100 g weight is about 2 g, then, comparing it with a heavier weight of 102 g or more will be discernible. For a weight of 200 g, the difference will have to be about 4 g, etc. The difference threshold ratio will be constant.

$$\frac{2}{100} = \frac{4}{200} = \frac{8}{400} = 0.02.$$

This is *Weber's Law,* after the man who discovered it a century ago or more. It is written generally as:

$$\frac{\Delta I}{I} = k$$

where: ΔI = the increase in stimulation,
I = the measure that was taken as the starting point,
k = a constant.

Table 5 gives the values of k for a number of sensory organs and there are marked differences that suggest that some senses are better for communication than others.

Table 5. Weber's Values for Different Senses

Sensory mode	k
Pitch of tone	1/333
Pressure on skin	1/77
Brightness	1/62
Weights	1/53
Amplitude of tone	1/11
Tasting salt solutions	1/5

2. *The second problem is the number of stimuli that can be identified at the same time*

This can be called a measure of "absolute judgement".

Interpretation

Once detected and identified, a signal will be interpreted, usually, and is given a meaning. The stimulus is related to the condition of an observer. This phase of perception is influenced by:

(a) instructions,
(b) previous experience,
(c) motives.

From the aforementioned, it will be apparent that perception depends upon the *aspects of people*. This is classified as the persistence factor.

5.6. The Isomorphic Energy Paradigm

In the book "Modern systems research for the behavioural scientist" by Walter Buckley, he elucidated how people qualify. For example, on page 186 it is written that "perception is the integration of sensory events" and further, on page 492, that "the significant environmental variety is transmitted via sensory and perceptual channels and decodings (*perception*) to the cortical centres, where, by selective criteria related to the physiological and/or other needs or drives (*motivation*), relevant parts of it are encoded and preserved as experience (*ability*) for varying periods of time".

From this it can be read that ability is a function of both perception and motivation and it can be expressed as:

$$\text{Ability} = (\text{Perception} \times \text{Motivation}). \qquad (2)$$

Assuming the approximation that equations (1) and (2) are linear, then, substituting (2) in (1) gives:

$$\text{Performance} = (\text{Perception} \times \text{Motivation}^2).$$

Earlier it was seen that industrial psychology formulated a number of complex units and, in this context, the following magnitudes were examined:

— Time (T)
— Motivation (M) (extension),
— Perception (P) (persistence).

Also the following assumptions can be made:

— action = $M.T^{-1}$
— impulse = $P.M.T^{-1}$
— ability = $P.M.T^{-2}$
— performance (power) = $P.M^2.T^{-3}$

The connections between these magnitudes are presented in Fig. 16.

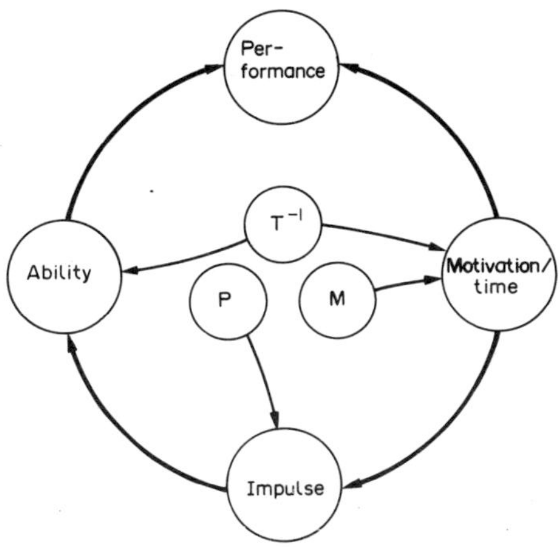

FIG. 16. Energy paradigm for industrial psychology.

5.7. Objective of the Isomorphic Energy Model

We can ask ourselves what part the isomorphic energy model can play in industrial psychology, because its implications are important for the personnel policies in industry. It could help giving insight into personnel selection, creating employment openings, improving ability (management development) and motivation policy (Maslow).

6.

The Isomorphic Energy Paradigm for Group Dynamics

"The welfare of the group has to be normative for the behaviour of each individual" (Plato).

Modern sociology has built up an understanding of social structures and processes. It deals with the relatively stable relationships between people and groups (social structures) and movements within these structures (social processes).

6.1. Structure

A social structure is a description of a fairly durable complex of social relationships and connections. Distinct groupings can be identified. Arbitrarily, the following examples are given: a theatre audience, a group of children at play, a political party, a military brigade, a set, a metropolis, an editorial board, or a team of accountants.

These groups have a structure which is broader than those found in physics, where the structures are factual and permanent, as a rule. A team of accountants, for instance, is not precisely perceptible, because its membership is changeable. The notion of structure, however, is usable as a means to understand and classify the great variety of social groups. A watcher of a football team in action may not observe its structure immediately, especially when players are substituted during the game. Thus, a social group can be called an open system (as opposed to a closed system in physics), because there is an interchange with its environment, although there is a continuity within it. The social structure of a school is another example, the membership changes all the time as new generations of children and teachers come and go; nevertheless, the overall characteristic principle and mutual relationships continue. In a student society, there is a regular flow of young students in and older students out, yet the same type of cliques and rivalries are retained. The same hierarchy is found each year and the same connections are maintained with other societies, alumni and the university board; therefore, it endures. Much the same thing holds for industrial organisations and commerce as a whole.

A number of divisions become apparent which are largely based on social relationships, and the following sub-divisions can be recognised:

(a) Groupings whose members have close social ties related to similar aims in life. Normally these are small groups and the members belong to the same family, team, platoon, neighbourhood, school, office or enterprise.
(b) Groupings whose members have loose social connections, or are "colleagues from choice", such as in a commercial bureaucracy, a political party, a church, a school, or the army.
(c) Groupings in which the members have little or no personal connection, may be from living in the same town or belonging to the same race, nationality, or age group.

These three groups can be named: a *group*, a *collective* and a *social class.*

6.2. Social Process

Activities which draw people to one another are called *interactions*. These interactions between associated people may be *co-operative* when they assist one another, or they may be *competitive* when they oppose each other.

In a competitive situation the members have different individual aims, whilst in a co-operative situation there is a common group aim.

In order to study competitive and co-operative behaviour, there are a number of relevant "games"; a classical example is "the prisoners' dilemma". It goes as follows: two prisoners have the choice of admitting or denying guilt. If they support each other and say that they are not guilty, each will get a light sentence (say one year in prison); however, if they both admit guilt, each will get a much longer sentence (ten years). If one admits guilt and the other denies, the first will be acquitted and the other will get a heavy sentence of fifteen years in prison. The situation is shown in Fig. 17.

		Prisoner A	
		Admitting (x)	Denying (y)
Prisoner B	Admitting (x)	Both get 10 years	B acquitted, A gets 15 years
	Denying (y)	A acquitted, B gets 15 years	Both get 1 year

The prisoners' dilemma

FIG. 17.

The dilemma is that if A could trust B to deny, A could by admitting be acquitted. Naturally, people try to maximise their own desires, and this could be attained by co-operation or opposition. When human behaviour is predictable, the game can be solved mathematically in order to find the most profit or least loss. However, most of the time people do not behave according to the best strategy.

In the example, when the prisoners co-operate and both deny, they will get the same minimal sentence, but if they compete, hoping that each can get an acquittal at the expense of the other, this will result in admittance by both and consequently a sentence of ten years for both. The purpose of this example could be the encouragement of people to work together; the situation shown in Fig. 18 leads, however, to a competitive behaviour.

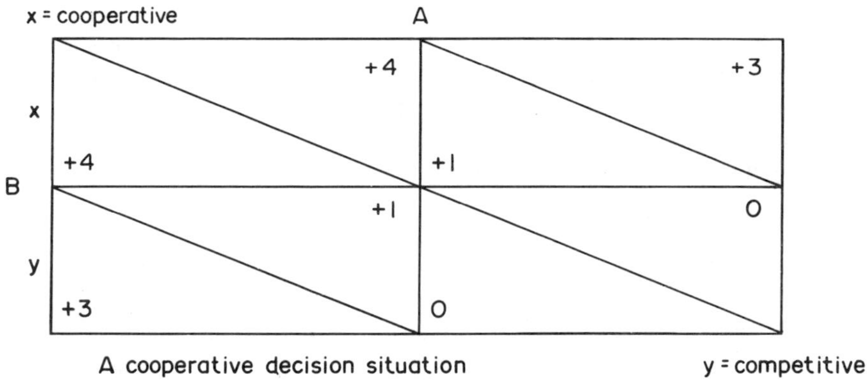

FIG. 18.

Co-operation always produces the best group result. When A chooses *x* he always gets a bigger profit than when he choses *y*, regardless of what B chooses.

A curious outcome of studying this game behaviour is that many players act competitively and choose *y* because they only want to get more than the other.

It can be concluded that people are competitive by nature; only when a co-operative action provides the best reward is the competitive urge suppressed. The instructions are important; if the players are told that the aim of the game is to be sure of getting as much profit as possible, the competitive behaviour dominates.

The communication between players has also a bearing on the results of a game; if they do not speak, the game will be competitive; when they speak to one another, it will be co-operative. Both of these situations have been studied experimentally, when it was found that co-operative groups showed the following characteristics:

1. Good co-ordination of all efforts.
2. Good productivity and quality of products.
3. Friendliness and favourable evaluation of the group and its members.

The performance of a group depends upon the way that the group objectives are formulated, especially whether the group objectives are in accordance with the objectives of the individuals. Even when members agree on the group objectives, tensions and conflicts may occur when implementing these objectives. Conflicting objectives interfere with the group processes and hinder its performance; also, the greater the competition, the less effective is the group as a whole. In an organisation, the differences between personal needs and the purpose of the organisation can lead to an *informal organisation*. When the needs of an individual are not satisfied, he becomes frustrated and looks for informal ways of getting satisfaction. Consequently, very few organisations function according to the policies that were originally conceived for them. Formal structures tend to be laid down by rules, but informal structures "grow" with familiarity, especially, when the executives deviate from the rules in order to achieve results.

These important aspects of group dynamics have been studied by Roethlisberger and Dickson (1939), and others.

The formal organisation, however, must not be underestimated, because no factory could operate successfully without it, no roads would be built, and no machines would function.

6.3. The Isomorphic Energy Paradigm

The football match, mentioned earlier, is an excellent example of the interaction process, because the action of one player (passing the ball) produces a reaction in another player (a shot at goal). A radio reporter, when describing the progression of interactions of a match, gives the expert listener an insight of the structure and functioning of the teams. The functioning of the teams is dependent on the rules of the game and the standards of play.

Values and Norms

Norms are conceptions about what should be done and they are the outcome of values. The function of norms can be made clearer using the notion order. Particularly, the ordering of interpersonal contacts plays an important role in social conduct.

E. G. Mesthene demonstrated that in a rapidly changing society also the values initially cherished by them change. This process has to be made conscious in order to give direction to a group's functioning. He proposed:

> "The emphasis will have to shift from values to valu-*ing*. For it is not particular familiar values as such that are valuable but the human ability to extract values from experience and to use and cherish them."

Consequently, groups of people have to develop for themselves skills to develop frameworks for values and norms under changing circumstances. M. Scheler and N. Hartmann described a value philosophy that was based upon the phenomenology of E. Husserl. They believed that it is possible to build up a hierarchy of values for the interactions between people in a group. When impulsive reactions are observed, they could form the framework for a set of norms.

The behaviour of society has to conform to a pattern of norms. Individuals have to know what is acceptable to the group and what is not. In Group Dynamics, the development of a scale of norms for a group is known as "norm setting".

The Connections between Interactions, Values and Norms

Social groups, processes and the conditions for optimal functioning are studied by sociologists. The constituent dimensions of group dynamics can be written as:

(1) time (T); (2) action (A) (extension); (3) value (V) (persistence).

It follows that:

(4) Interaction per time unit $= A.T^{-1}$
(5) Group dynamic impulse $= V.A.T^{-1}$
(6) Norm setting $= V.A.T^{-2}$
(7) Group dynamics power $= V.A^2.T^{-3}$

The connections between these dimensions are illustrated in Figure 19.

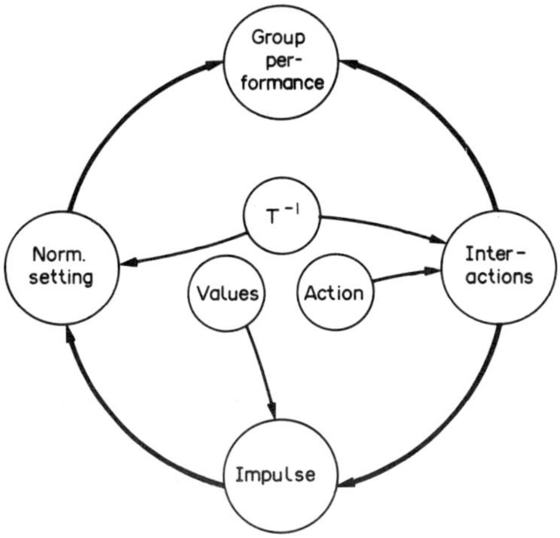

FIG. 19. Energy paradigm for industrial psychology.

7.

The Isomorphic Energy Paradigm for Industrial Management

"We look forward and we look backward, and we yearn for what is not there" (Plato).

7.1 Introduction

Organisation and management studies are concerned with integrated theory of organisations (structure) and control (functioning). A few functional aspects are presented here, leaving out the structural aspect, for the isomorphic energy paradigm is concerned with the functional aspect within structures.

With the aid of such a paradigm, industrial situations can be explained better and directions given for making their structural framework. The isomorphic energy model can show how the parts of a system are related to each other. Generally, it states that power is energy per time unit and that the form of energy comprises three constituents, namely, time, a persistence and an extension.

Management of events requires managerial energy in order to enable influencing them in such a way that a desired state will be attained.

The constituents are:

— time,
— implementation (extension),
— control (persistence).

The derived dimensions are:

— implementation per time unit,
— managerial impulse,
— measure of planning,
— managerial power.

7.2. Managerial Organisation

A managerial organisation is necessary for instructing an industrial component or production module. These actions depend upon the objective, the starting phase, the daily events and the plans.

The following activities comprise managerial organisations:

— Planning — all the activities that lead to the achievement of a decision.
— Implementation — the implementation of a plan which will be realised through management activities, i.e. transforming a decision into a fact.
— Control — the recording in a report of the actions connected with the progress of a plan.

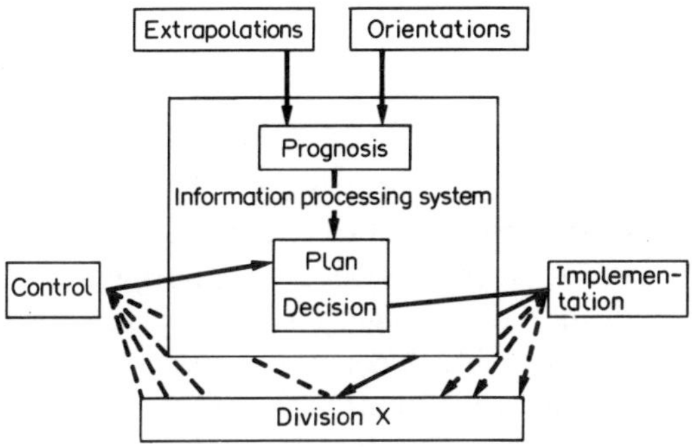

FIG. 20. Information processing unit.

Planning

Planning includes all the activities that lead to the achievement of a decision. The work activities are recognised by the following:

— the collecting of reports, extrapolations and orientations,
— the drafting or adapting of necessary prognoses,
— the drafting of a first plan,
— the analysing of the first plan,
— the testing of the first plan.

Reports, extrapolations and orientations are combined and evaluated in order to arrive at the prognoses, which are describing the expected progress as good as possible.

Prognoses are difficult to make and exploring many other plans is necessary in order to avoid creating conflicting situations. Care must be taken to enhance the effectiveness of prognoses.

Before arriving at a decision, it is necessary to make a number of preliminary plans. The plan that stands the test turns into a decision. If we talk about improving the quality of planning, i.e. always making better descriptions, we have a scale in mind of a number of planning techniques for improving the practical success of plans. The following can be mentioned here:

↑
— etc.
— double entry
— SCRAPP
— PERT
— Gant Bar Charts

We attribute a certain value P to the level of planning.

Implementation

Implementation (U) means execution of a decision, and speed of management means the total number of actions per unit time. The present economic development (see the isomorphic energy paradigm in the macro-economy) requires that the industry reacts more quickly on the world around us. This says, in fact, that if the events in a certain module were spread over five months, four months should be aimed at, which involves more actions per unit time. This can be expressed as $I = U.T^{-1}$.

Control

The activities of a management organisation begin with a certain situation which has to be changed by a series of events into a final situation. For that purpose, the condition of the module must be known. The results of management actions are made known in reports.

A report always refers to a previous period of time. What concerns a report is always the choice between a message that explains the new situation and a message that explains its deviation from the original situation. Control finds its origins in responsibility for borrowed capital. The limiting conditions which are imposed on the production process by present-day society are always increasing. The amount of control which is needed in order to meet these conditions is increasing too. However, this hinders a quick reaction to the complicated outside world; consequently, resources are required and acquired in the form of quick processing of information, reporting exceptions, trend analysis, selectivity and so on, in order to keep the amount of control as small as possible.

Also, for the control dimension, we can visualise a scale wherewith the increasing amount of control can be expressed. We attribute a certain value C to the amount of control.

7.3. Management Power

In a management organisation the planning function is related to the management actions per unit of time. A higher planning level makes possible more management actions per unit of time. This relation can be written as

$$› P \rightarrow › U.T^{-1} \quad (1)$$

Without knowing the conditions of the module by the control function, the planning function cannot be executed. It is desirable to use as little management energy on control as possible. Applying the selectivity principle to the control function appears to reduce the amount of control per unit of time, thereby facilitating the increase of the number of management actions per unit of time.

We write:

$$‹ C.T^{-1} \rightarrow › U.T^{-1} \quad (2)$$

With equations (1) and (2), planning, control and management actions can be united in a relationship which is written as:

$$\frac{\mathrm{P}}{\mathrm{C.T^{-1}}} = U.T^{-1} \text{ (or equally as } P = C.U.T^{-2}\text{)} \quad (3)$$

If a management organisation functions when such elements are recognised, the power of management (*MP*) can be equated:

$$MP = P.U.T^{-1} = \mathrm{C.U^2}.T^{-3} \quad (4)$$

The management impulse (*MI*) and the quantity of management energy (*ME*) which is used when the plan is executed can be expressed as:

$$MI = C.U.T^{-1} \quad (5)$$

$$ME = P.U \quad (6)$$

The connections between these constituents and resultant dimensions can be illustrated as shown in Fig. 21.

Included with the management power is the power for obtaining certain degrees of planning and amounts of management actions per time unit. An example might explain a few points.

For a given managerial organisation, the values for *P*, *C* and *I* are 80, 4 and 20, respectively; then $MP_1 = 1600$. If the number of management actions are to be doubled whilst the amount of control remains the same, the values become: 160, 4 and 40 respectively.

Consequently, MP_2 becomes 6400, which is four times as great as MP_1.

The demand for quicker reactions to the outside world is mainly a question of reducing supply time, however on the condition of earning capacity. This holds in so much as to keep stocks as low as possible which demands a high degree of planning and scheduling. The weighing, on one side, of costs of an efficient information system and, on the other side, keeping the required stocks, is based on a well-known economic principle. The criterion for evaluating a management system must not be the stock level; this will lead to combating only symptoms and other side-effects.

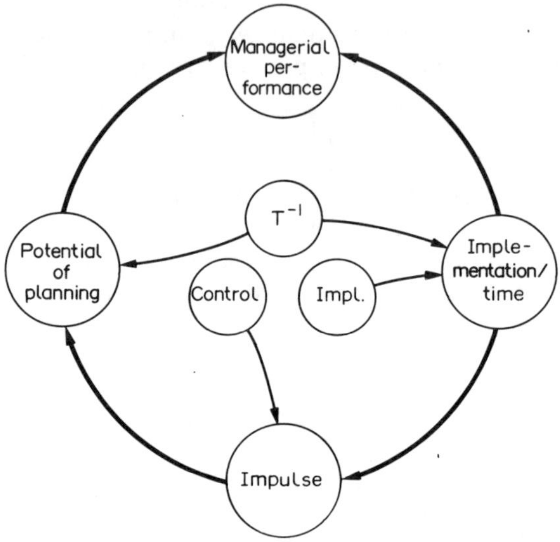

FIG. 21. Energy paradigm for industrial management.

The criterion should be the whole of the business efforts, from start to finish. If it is required to reduce the supply times, then it will have consequences for all the modules that comprise the whole.

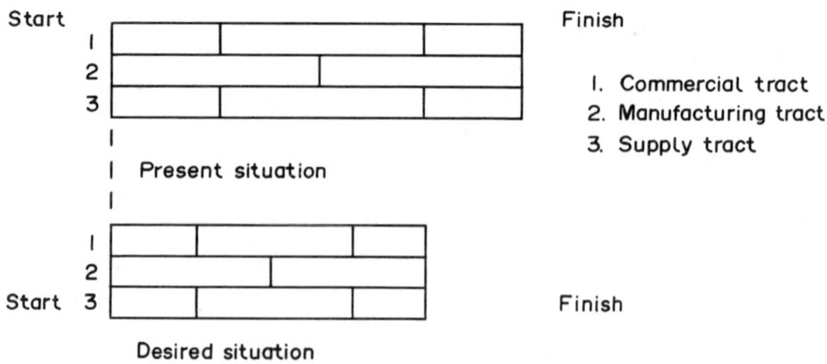

FIG. 22. Reducing supply time.

In order to attain the supply time 2, the number of management actions per time unit will have to be increased; thus, more managerial power is needed. The quality of prognoses must be improved, and planning activities and techniques developed (›*P*).

The control function should be improved by applying the selectivity principle and by developing automatic information processing systems (‹*C*).

The time spans for the modules must relate accurately to each other. An increase in managerial power in a single module will not at all affect the productivity as a whole.

Finally, the management units should grow into an integrated system.

In another example it was advised to introduce a more frequent ordering procedure in order to attain a reduced (*V*) inventory level (see Fig. 23).

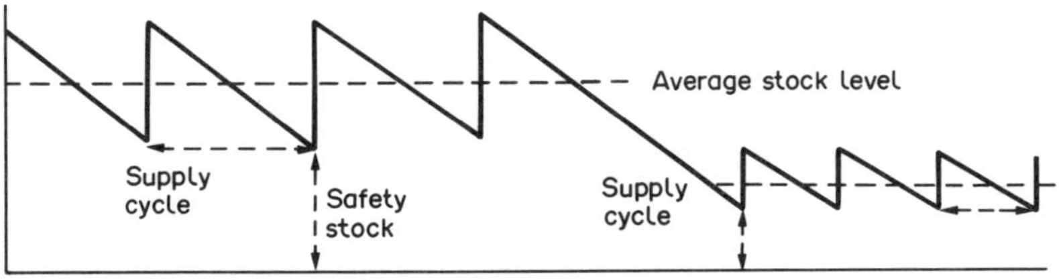

FIG. 23. Frequency of ordering procedures.

Indeed, could this be possible if at the same time the production were adapted to the new procedure? Since the planning of stocks and scheduling of production were not tuned in, a situation arises indicated as V_1 in Fig. 24.

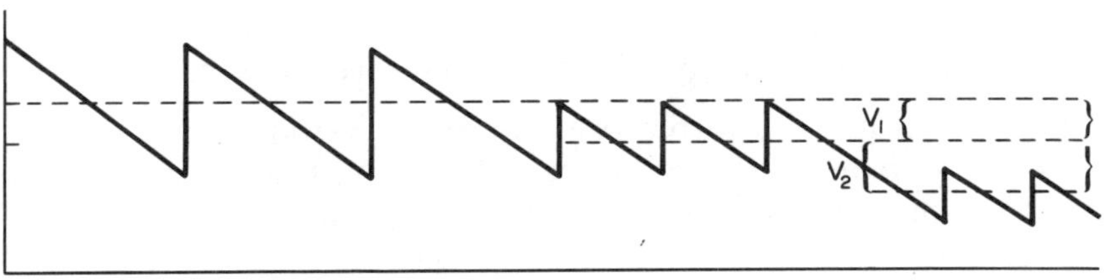

FIG. 24. The integrated system.

Only at a later stage, after tuning in the manufacturing process, will the total benefit of $V_1 + V_2$ result.

7.4. Sketch of a Problem Identification

A company manufactures 2000 catalogue items of electrical equipment.

Average turnover time = 15 months.
Average work-in-progress = 800 orders.
Average number of orders concluded by the commercial department = 4 per day.

Average number of quotations handled per day = 20.
Various production lines.

Actual structure and work methods assumed

1. Mechanical workshop — 500 jobs per week.
2. Materials obtained internally.
3. Materials supplied from outside.
4. Assembly department — job descriptions.
5. Testing department — job descriptions.
6. Stores — issues range from 40,000 to 2,000,000 items per year.

Problem definition: to reduce the turnover time from 15 months to 12 with improved information processing.

(a) *Planning.* The planning system used Gant bar charts. Much time was consumed in planning meetings. Next a new system based upon PERT was introduced. Each order accepted by the commercial department was incorporated in a PERT plan.

The inter-relationship of project structures to the production system could not be clearly distinguished, but insight into this matrix problem was really an issue of making decisions.

Now, an advanced planning system has been introduced in which 800 PERT plans are matched with the production structure by a computer. This procedure results in the orders and work tasks for activities 1 to 6 (1500 management actions per week).

(b) *Reporting.* Actually, if everything runs according to the plans, no reporting is required. However, this ideal can never be achieved due to breakdowns:

— technical breakdown; material breakages, machinery and equipment breakages, etc.
— organisational breakdowns; sickness, absenteeism, waiting times, etc.

To make plans, the current state of affairs of production has to be known. This requires reports. Reports hardly ever describe the present state, only previous situations. It occurred frequently that the commercial department had to deal with two-months delays in reporting. Disturbances compelled the management organisation to revise their plans, but all attention is required to handle the quotations of 20 per day. As little energy as possible has to be used for revising plans, and the aim should be to reduce reporting to short summaries and to deliver them as quickly as possible. Eventually, a daily control board has to keep people up-to-date. This results into a substantial reduction in reaction time.

(c) *Increasing Managerial Power*

1. Defining as accurately as possible the future (planning ›).
2. Minimise reaction time (reporting ‹).
3. Minimise turnover time = maximise management actions per time unit.

All three functions can be measured.

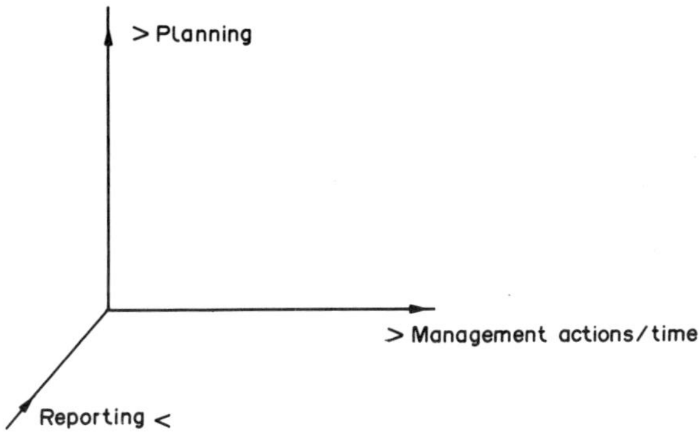

FIG. 25. Basic functions of management.

The Situation in Business Life

We see every day that management is shackled by slow reporting and this affects the level of planning and the number of management actions per unit of time (the turnover time increases). This leads to the saying "Daily routine drives out long-term planning." One does not yet recognise that the increase of complexity in business life requires improving management-power drastically.

With increasing competition, the problem lies particularly in the delivery times, and consequently in the improvement of managerial power.

The KIVI has published a report in which it is calculated that 2000 planners are needed in Dutch business life and that training planners should have prime importance. This is, as yet, a one-sided approach and the problem of understanding managerial power must be tackled, with management training coming first.

Figure 26 illustrates managerial power. A certain managerial power, MP_1, is characteristic of P_1 and I_1. If I_1 should proceed towards I_2, whilst maintaining the same reaction time, the managerial power should increase to MP_2 by increasing P_1 to P_2.

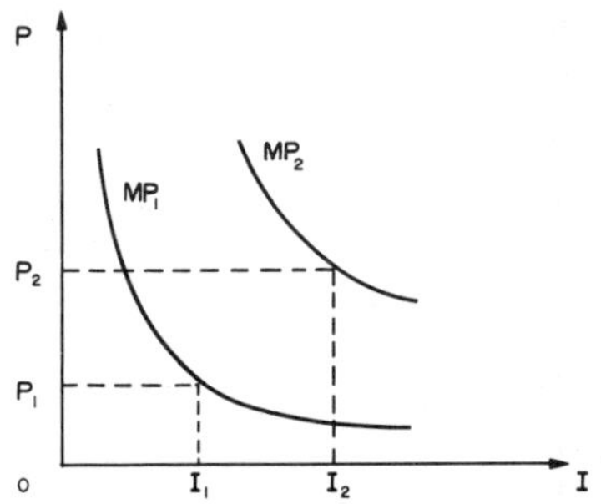

FIG. 26. Diagram for managerial power.

8.

The Isomorphic Energy Paradigm for the Macro-economy

"We must cultivate our garden" (Candide).

8.1. Introduction

The characteristic of neoclassical economy is the uniting of production with a number of production factors that can be explained by the functional equation $X = f(K.A)$, where X symbolises the Gross National Product, K the capital stock and A is the amount of work expressed in man-hours. The related theory concerns the form of this function, on the one hand, and the implications of its different forms for a number of economic phenomena, on the other hand.

The capital stock is taken here as a statistical magnitude, but already Henschel argued that the *technical qualification* of a population also determines the magnitude of the capital stock. Jan Tinbergen also pointed in his research priorities at "the potentiality of education as a means for increasing people's productive capacities".

The energy paradigm offers an opportunity to relate technical qualification to the investment in capital stock. Also, consideration will be given to replacing "work performance in man-hours" by a work extension per unit of time.

8.2. Laboriousness (Extension per Time Unit)

The available labour makes a difference to the quantity of work per time unit. For example, in the port of Rotterdam, a gang of dockers may unload a certain amount of goods from a ship in one day, but in Bahia, a gang of the same size may require three days for the same amount of goods. The laboriousness of the dockers in Rotterdam seems higher than that of those in Bahia. The difference is caused by a number of factors, including motivation, work disciplines, rest times, and such like. In order to find a relationship with GNP, a comparison between the average work done per unit of time on a year basis in The Netherlands and that done in Brazil will have to be made. If, for instance, for every 300 work days per year in Brazil there are 250 in The Netherlands, the laboriousness in Brazil could still be higher. The comparison is shown diagrammatically in Fig. 27, where the solid lines represent The Netherlands and the broken lines Brazil. Laboriousness is taken as the work done per time unit and it is expressed as $L.T^{-1}$.

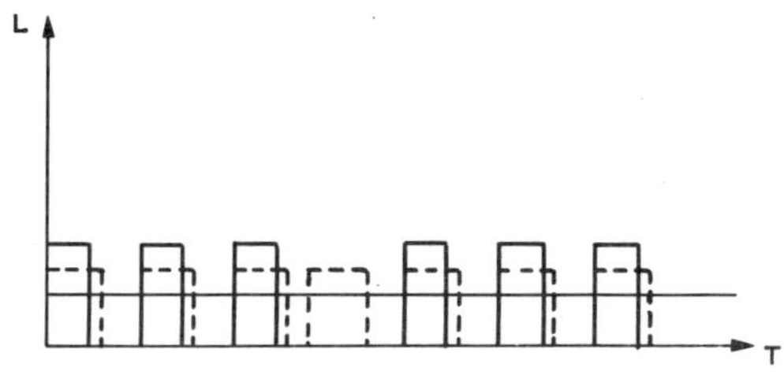

FIG. 27. Labour per time unit.

8.3. Technological Qualification (Persistence)

The technological qualification of a nation is of great importance to achieve a bulk of capital stock. However, the type of education and schooling, improvements in health, research, replacement of products and production processes, scale of production, market organisation, mobility and objectivity of management, all have an influence on technological development.

Already, many attempts have been made to find starting-points for measuring technological qualification, such as the increase in expenditure on scientific investigation, as shown in the graphs of D. van Duyne in *Maatschappij in toekomstperspektief*. Also annual numbers of physics publications, number of qualified graduates per year, etc.

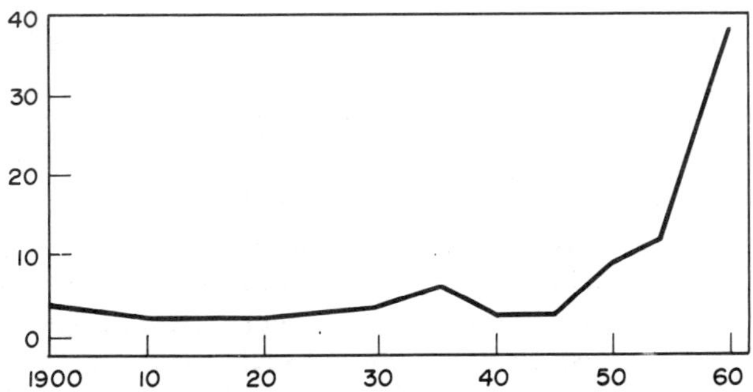

FIG.28. Annual numbers of physics publications, in thousands.

Theoretically, a scale of technological qualification can be produced and it will be represented with the value H.

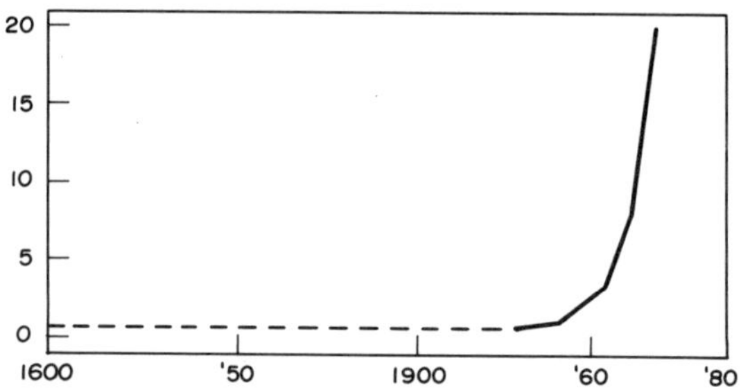

FIG. 29. Annual expenditure on research and development in the United States, in billions of dollars.

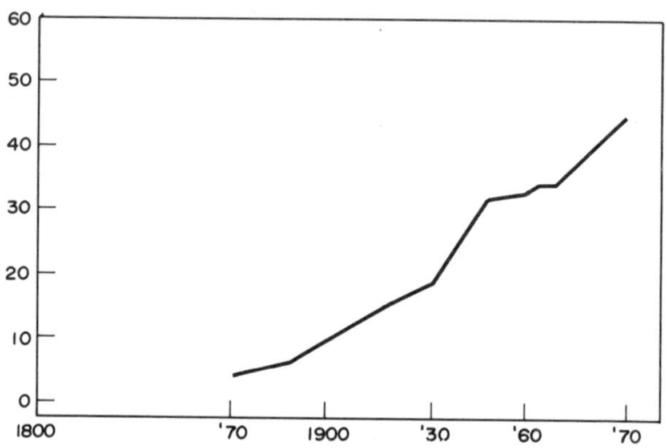

FIG. 30. Qualified graduates from higher education in the Netherlands per 10,000 of the 25–29 age-group.

8.4. Economic Impulse

Economic impulse will be expressed as the product of technological development and laboriousness. The average economic impulse of a nation is determining the magnitude of capital stock; for example, the experience of the Marshall Aid scheme can be contrasted with the help given to developing countries.

After the Second World War, a greater part of the European production capacity was destroyed, but it was quickly built up again with the help of dollars, because the technological qualification and manufacturing experience in Europe was still present on a high level, together with a high degree of laboriousness. Aid for the developing countries in the form of funds for building up capital stock, on the contrary, has little effect. The build-up has to be gradual; that is in direct relation to the level of technological qualification and the national laboriousness.

The equation is:

$$I = \text{economic impulse} = H.L.T^{-1}.$$

8.5. Capital Stock

The total capital stock per employee is assumed to be equivalent to the machinery available to the average worker. As such, it considers equipment with a life of more than one year, including factory installations, electrical apparatus, transport and distribution facilities.

The gross investment includes replacement and the so-called major maintenance and improvements which can lengthen the life of production facilities. In general, the figures for expenditures related to the procurement of capital stock for production have to be taken into account. Calculations from The Netherlands Economic Institute in US dollars for a number of countries will serve to show national differences:

USA	\$23,605
Sweden	\$20,389
Holland	\$14,526
France	\$14,221
Belgium	\$12,493

We see the volume of capital goods supplied as a time differential of the economic impulse, $K = I.T^{-1} = H.L.T^{-2}$.

8.6. Gross National Product

Prosperity is written as the Gross National Product in which is calculated the total output of goods and services supplied. The GNP per employee is the total GNP divided by the number of employees. It is the result of the co-operation of labour and capital stock. The relationship can be expressed as: $\text{GNP} = K.L.T^{-1}$.

This relationship will be illustrated by a diagram (see Fig. 31) which expresses the economic power of a national production economy.

From this diagram it can be seen that a GNP of the order GNP_1 can be obtained with a capital stock of C_1 and a laboriousness L_1; similarly, for GNP_2 with C_2 and L_2. Increases in these components will produce a greater GNP, just as GNP_2 is greater than GNP_1; whilst GNP_3 is greater than GNP_2. Measuring the distances along OW shows that the differences between the GNP values are equal. For the countries studied by the NEI, the following values are obtained:

USA	$9,217
Sweden	$6,285
Holland	$4,878
France	$4,734
Belgium	$4,622

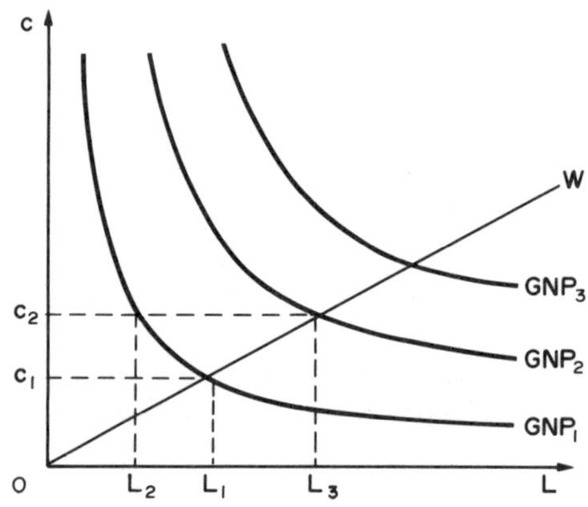

FIG. 31. Diagram expressing economic power.

8.7. Calculation of Some Economic Ratios

Only statistical figures of three of the dimensions of the paradigm are available up till now; namely, gross national product, investment in capital goods and time. For the latter, we take the average number of the hours that are worked in a particular country. In The Netherlands, we can take this figure as 2000 hours. Using these values and the equations enables the rest of the known ratios to be found. These ratios indicate at least the order of magnitude for the macro-economic components in the previous countries.

TABLE 6

	GNP	CAP	Impulse (100,000 $ hours)	Labor-iousness $.c	Techn. qual.	Labour	Time
USA	$9,217	$23,605	425	39.0	10,9	$702	1800
Sweden	$6,285	$20,389	367	30.8	11,8	$554	1800
Netherlands	$4,878	$14,526	290	33.6	8,5	$672	2000
France	$4,734	$14,221	284	33.3	8,6	$666	2000
Belgium	$4,622	$12,493	262	36.9	7,1	$775	2100

The connections between the components of the US macro-economy are shown in Figure 32.

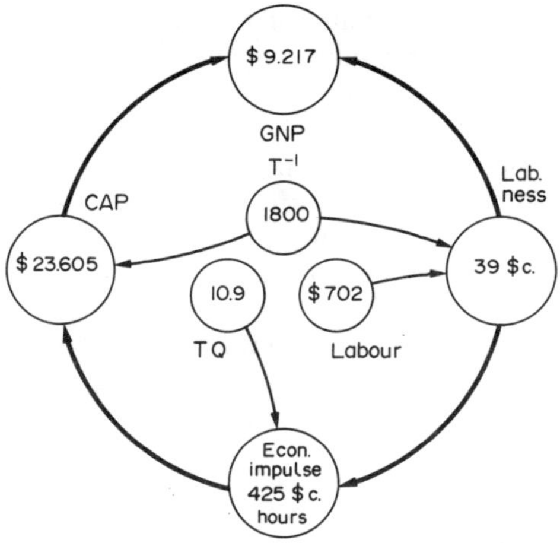

FIG. 32. Energy paradigm for the macro-economy.

From this survey it appears that:

The greatest contribution by work to the economic process is in Belgium.
The least contribution by work to the economic process is in Sweden.
The highest technical qualification is in Sweden.
The lowest technical qualification is in Belgium.
The largest laboriousness (work per time unit) is in the USA.
The smallest laboriousness (work per time unit) is in Sweden.
The Netherlands occupies a mean position in each instance.

When we put GNP, capital stock and laboriousness into a diagram, we get the following results (Fig. 33).

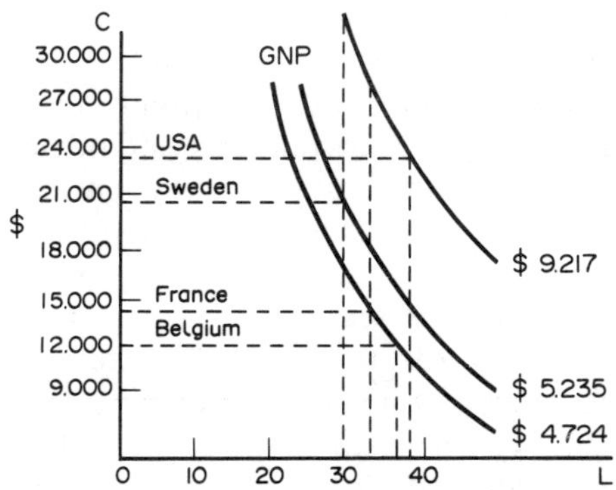

FIG. 33. Isoquadrates of GNP for a number of countries.

From this diagram one can read off the amount of investment in capital stock for France and Sweden in order to see how much they must invest per head of the working population in order to enjoy the same prosperity as the USA. Keeping laboriousness constant in these countries, France should invest \$13,000 and Sweden \$10,000 to get an increase of \$4,500 and \$3,000 GNP respectively. With an increase of 30% in the capital stock investment over Sweden it will be possible for France to get a 100% increase in GNP, yet the Swedish GNP increases only 46%. This difference is largely due to the different laboriousness values which are \$c 33,3 in France and \$c 30,8 in Sweden.

8.8. The Aim of the Isomorphic Energy Model

We must ask ourselves where the isomorphic energy model fits into the economy. It finds its place next to economic statistics and short trade cycles. The aim of economic statistics is to try to balance the national economy, presuming that population, capital stock and technological qualification remain constant. The theory of short trade cycles tries to explain these movements, such as those caused by temporary overproduction in certain industrial branches. In these models the supposition assumes also constant populations, constant investments and a fixed state of technical expertise.

The isomorphic energy model is concerned with the investigation of economic developments over periods of decades and such a theory may not overlook laboriousness, technical qualification and the capital growth, etc. In fact, these are the particular objectives for its method of analysis. The central problem, for the sake of brevity, can be put by the following question: How do production, employment and living standards change under the influence of changing laboriousness, technological qualification and capital stock? It involves trying to get an insight into the economic process and then determining the measure of its sensitivity for control purposes.

Restrictions

1. One should realise that the isomorphic energy model is not concerned with short-term fluctuations. This restriction is necessary in order to simplify the analysis.
2. Moreover, long cycles are left out of consideration, because their effect is mainly on price fluctuations and their influence on production, employment and capital investment, even fluctuations in wages, are secondary. This is not to say that they are irrelevant, but that the fundamental problem depends upon the discovery of the causes of continued increases in the named variables. Our analysis is mainly directed towards the decisive determinants of economic growth and the way that it can be influenced.
3. This approximation is perhaps more pertinent to the third restriction which is the exclusion of monetary aspects. These are of special interest for short-term movements and mainly for explaining price fluctuations, wages and interest costs. In the case of long-term movements, these last fluctuations can be separated almost entirely from the real dimensions of the economy.
4. Finally, the total economy is being considered without the complication of individual markets and goods.

9.

An Actual Situation in which the Managerial, Individual, Group Dynamic and Economic Powers are Distinguished

A number of American businesses have introduced so-called "management services rooms", where one can find all the information relevant to management, such as:

— progress of prices of raw materials,
— stock exchange quotations of competitors,
— planned and realised purchases,
— investment progress reports,
— work in hand,
— costs versus budgets,
— and so forth.

These items are displayed on wall charts assisted by computer programs. The frequency of the forecasts depend on the interest to management. Generally, sales returns are posted daily, whilst profit and loss reports are prepared monthly.

Next to this management information there have been prepared operation research models with computer programs, which help management to reach conclusions and make decisions.

Examples include:

— what will be the profits if the prices are reduced by $x\%$, taking into account the expected increase in turnover and a lower cost price?
— what are the consequences for orders that have been accepted already, if a substantial rush order is being accepted?
— which factory must be allocated a new product, taking into account the production and distribution costs, etc.?

9.1. Use of Management Services Room

Management services room is connected with the management offices and replaces all routine information. No statistics for individual use are distributed. The management service room:

— procures for management recent and relevant information about the macro-economy.
— procures for management recent and relevant business information.
— promotes communications between line and staff (group dynamics), with reference to planning and production.
— promotes individual performance.

There the following meetings are being arranged:

— of managers each Monday with reference to the state of affairs (45 minutes).
— staff discussions each week to consider results (30 minutes).
— top-management get together each month to discuss progress (2 hours).

One would expect that this framework would lead to a more centralised management, but in practice it is not the case. The useful information frees management from involvement with daily routines and they have more time for important matters and long-term problems.

The framework develops a new sort of management that is more energetic and shows a higher measure of perception and motivation. This keeps managers on the right line for adapting themselves quickly to social and market changes and to internal events, which is necessary in our fast developing society.

Alertness requires firstly a good co-operation based upon motivation which can be obtained only if those involved see and understand better what goes on outside their own departments. The full development occurs in three stages:

1. Collecting data and procuring information in a way that is appropriate to the needs of management so that it leads to a greater ability to survey and manage.
2. The greater ability to survey activities leads to the promotion of policies for optimalising a number of economic aspects.
3. This quantitative approach includes the preparation of better decision models for policy-making.

10.

Identification

"Tat-Tvam-Asi" (Bhagavadgita)

10.1. Introduction

In line with the view of Johnson, Kast and Rosenzweig in *The Theory and Management of Systems*, one of the most urgent questions for scientific research in modern society is the problem of identification of man in relation to the technological production system. The crux of the problem was seen by the psychologist Kennith Kenniston as "the inability of a man to fulfil his egocentric needs in an environment from which he is separated by the demands of a technological age". Alienation becomes a vogue word and the guildsman is exalted because of his professional skill. On the other hand, the modern operator sees in a distillation column only a number of measuring instruments and absolutely nothing of the product itself. The environment becomes filthy and one tries to escape from it, either through seclusion or by going sick.

The time of the guildsman is not likely to come again, and the question may well be: How can the personal ego fit into modern society? At least, there are two sides to be understood; on one side, mankind wants to display its identity and society must allow it freedom so that one can choose or develop his own structural form. On the other side, a binding element will be necessary so that "reality" becomes communicable.

The latter can be found in energetica, which gives the insight into fundamental aspects of reality, finding that all aspects of reality are forms of energy. The problem of identification can be put in the following way: how can I recognise the many forms of energy around me and how can I feel myself as one with them? In the first instance, it must be recognised that energy plays the greatest overall part, for it comprises personal energised power (power is energy per time unit). This recognition is divided into two parts, physical and psychic.

Physically, the possibilities include exercising physical games. In my opinion, in practising sports, the components of energy should be mentally identified. As an explanation, this can be easily experienced physically. What is physical power?

For instance, with the aid of a damper (see Fig. 34), in which a piston can be moved back and forth, the volumetric dimensions can be physically experienced.

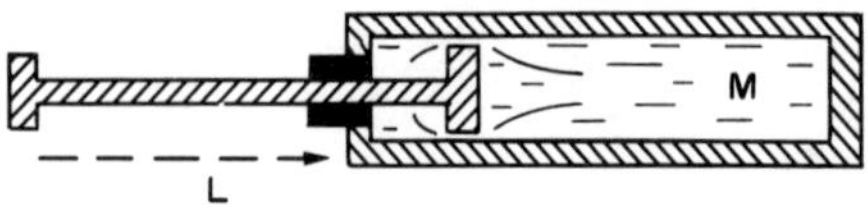

FIG. 34. A damper

Pressure resistance (mass replacement per second) $= M.T^{-1}$ (kg/s)
Action (movement of the piston per second) $= L.T^{-1}$ (m/s)
Force $= (M.T^{-1}).(L.T^{-1})$ (kg . m/s^2)
Power $=$ *force* $\times$ *action* $= ML^2.T^{-3}$ (kgf . m/s)

Getting a notion of each specific dimension will become possible by combining variable magnitudes. For that purpose, one can measure his own forces with two different dampers, by moving the pistons at different speeds.

Countless examples can show the agreement between the energy forms, but we limit ourselves here to the presentation of the well-known electrical circuit with a battery, lamp, ammeter and wattmeter.

Components		*Symbol*
lamp ≈ resistance	R	Ω
ammeter ≈ current flow ≈ *action*	I	A
battery ≈ voltage ≈ tension *force*	Emf	V
wattmeter ≈ power	P	W

force ≈ tension $= R \times I$
watt = voltage × current $W = V \times A$
power = *force* × *action*

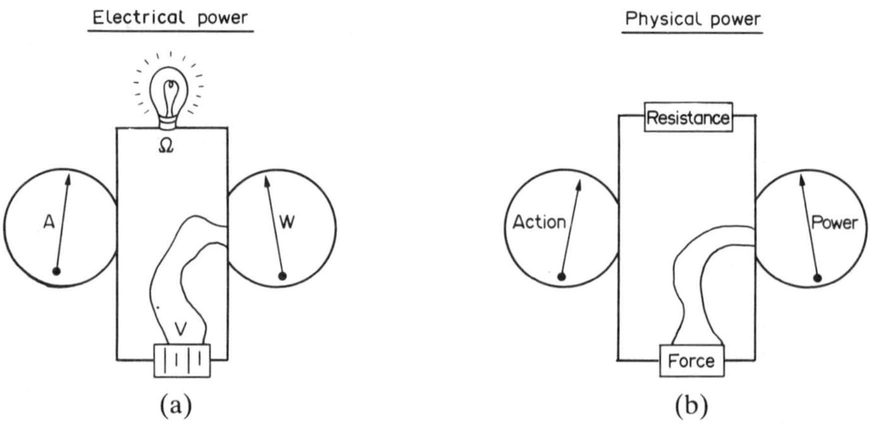

FIG. 35. (a) Electrical circuit; (b) Mechanical circuit.

Personal thinking requires energy as a basic input too, and we can distinguish three categories:

1. An energy shortage.
2. An energy balance.
3. An energy surplus.

An energy shortage will result in regression of the structures which were developed in an earlier learning process. An energy balance can be considered to be a normal situation in a health relationship between man and society, whilst an energy surplus supplies the dynamic structures which make it possible for society to continue. First we will set out how the energy shortage manifests itself in the social behaviour and thinking of individuals.

In his study of patients with brain damage, Goldstein defined a number of phenomena which are summarised as follows:

1. Following their injuries, a certain mental capacity or area of operations was never eliminated separately. Always, all activity areas were affected to some extent at the same time, although the intensity of the disorders for each of the different kinds of physical activity could differ.
2. A certain area of mental activities, such as reading, speaking or playing the piano, was never lost completely; there always remained some elementary ability to use this faculty.
3. The disturbances which a patient displayed with respect to the different areas of physical activity were all of the same nature. The different symptoms can be conceived as manifestations of one and the same fundamental change.
4. This fundamental change is demonstrated whenever the patient fails to perform his operations, if he can only get his bearing with the help of fantasy. If, on the other hand, results were to be obtained through factual behaviour that referred to material that was present in a tangible manner, the patient behaved adequately. From these characteristic symptoms of sickness, it appears that the illness had a distinct epistemological character. One can hardly expect otherwise in cases of brain damage, but exactly how the cognitive structure through reduced functioning of the organic base is affected is very interesting epistemologically.

The symptoms shown by patients who have suffered brain damage can be characterised epistemologically as follows: pathological cognitive structures are characterised by a loss of the power of abstraction and a related decreased ability to discriminate. By the power of abstraction, we mean the possibility of performing a mental activity out of context. One sees this in patients whose isolated actions are being more affected than general ones; it is impossible for them to locate a particular item in a picture, although they recognise it globally.

The steps of disintegration follow degrees of brain damage and they suggest a hierarchy for epistemological performance, or a structure of mental operations with the most vague, or global, at the base. Brain damage eliminates an energy surplus in the organism, that was reserved for the most abstract and refined cognitive structures. The elimination of part of the brain cortex activities reduces the structural power of the man by depriving him of the most efficient means of personal action. In a definitive environment, the patient does not use his power of abstraction: the situation makes him aware of what to do and how to do it. He does not need to structure his behaviour pattern in advance, nor does he need to visualise the scene in which to place the objects he wants to deal with in an abstract form, like writing a treatise, constructing an apparatus, or painting a picture. The setting for a given factual structure asks for less energy of the cognitive structures than visualising an imaginary environment.

The following test presents an opportunity to recognise "mental power" in the form of energy. A well-known psychological test, the so-called speed and skilfulness test, will be used here. With the aid of a computer program,[1] letter and number combinations are being projected respectively on a display screen of a PDP 11/40:

cu	uc	us	uu	cc
la	81	18	8a	a8
.	.	.	.	.
.	.	.	.	.
.	.	.	.	.

etc

One of the combinations (in example us and la) has no relation to the other combinations and the test person has to single out this one by using a light-pen. There are four series on display:

1st series of 30 groups which are shown for 6 seconds per group
2nd series of 36 groups which are shown for 5 seconds per group
3rd series of 45 groups which are shown for 4 seconds per group
4th series of 60 groups which are shown for 3 seconds per group

The correct answers are recorded and the number of correct answers per series are typed out.

[1] See Appendix 83/85.

It needs no explaining that the number of correct answers decline according to the increase of the speed of display. We can now define:

v = display speed: number of groups per minute,
t = tension: number of correct answers per standard number of groups (in this case, per 10 groups),
s = selectivity: it appears that the speed is proportional to tension so that we can write $t = s.v$, when s is a value for selectivity,
P = power: a product of tension (force) and speed (action): $P = t.v = s.v^2$

We assume that for the duration of this test the "mental power" for this activity remains constant.

The results from a number of tests gave the following average figures:

speed 1	= 10	speed 2	= 20
tension 1	= 6	tension 2	= 3
selectivity 1	= 0.6	selectivity 2	= 0.15
power	= 60	power	= 60

Using the well-known diagram of an electrical circuit, these results can be illustrated (Fig. 36).

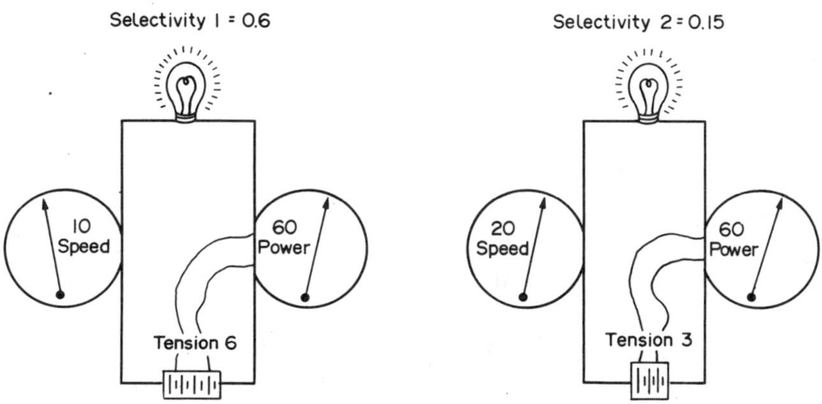

FIG. 36. Mental circuit.

From this experiment, one can easily identify one self with other models such as the industrial psychology model, the management model, etc. (see Appendix 78).

N.B. The Multiple Cue Probability Learning Models (MCPL) experimented with dimensions such as Performance, Knowledge and Control (Hammond and Summers, 1972: Cognitive Control).

10.2. Length, Area and Volume Structures

The general energy system shows that the development of energy forms expands from simple to increasingly more complex systems in rank order according to the progression:

length — area — volume dimensions

It is known that:

in mechanics, the dimension is L^1 (length)
in areal tension, the dimension is L^2 (surface)
in hydraulics, the dimension is L^3 (volume)

The plant kingdom can be developed similarly:

single-celled life forms L^1
seaweeds (all cells in a flat layer) L^2
higher organisms (e.g. the formation of transportation channels) L^3

The development of nervous system uses the same three dimensions:

neuron L^1
nerve bundles L^2
brain L^3

Business management develops in the following way:

simple bookkeeping L^1
business information L^2
integrated business information L^3

(see Appendix 82)

Also, in social organisations the three-dimensional control information systems must develop as a condition to the block formation such as prognosticated by Adams (see Introduction).

10.3. Conclusion

The Sun, Earth, Society system forms one energy complex. As long as the Sun shines, life on Earth will be possible. Our knowledge has gradually increased in order to create for people of our "Global Village" (MacLuhan) the conditions for a rational existence. Therefore, the upbringing for future generations must take the following basic direction:

1. *Identification:* that they learn in sport and experiment to identify themselves with "Nature", i.e. energies, inorganic and organic systems, production and social systems.
2. *Synergetica:* we must learn that working with good team spirit, can have synergetic effects. Through (1) and (2) the possibilities become greater for each individual and for cultural groups to develop talents.

3. *Organisations:* for the development of national blocks, more advanced communications and control systems, as well as organisation structures will have to be developed.

Man is composed of 10^{14} cells and each cell is regarded as an individual unit of life. This unit is regulated by a nucleus, but the brain co-ordinates all these individual cells. Should not we be able then to design a world structure in order to co-ordinate the world's population of 4×10^9 individuals?

Life is imaginative. On one theme, that of *force* × *action*, are composed 1,000,000 types of plant, 700,000 types of insect and yet another 1,000,000 types of animal. What a wealth of structures. A dead nettle is certainly not an oak, but both manage their water supplies on the same law: *force* × *action* (suction pressure × water flow).

For psychologists, sociologists, economists and organisers, there is plenty of work to do. "Il faut cultiver notre jardin" said Candide two hundred years ago.

11.

Closing Remarks

"L'énergie se faisant présente" (Teilhard de Chardin).

Through the ages, people have thought about finding a synthetic unity for multiplicity, and answers have been proposed that were based upon the information available at the time. As they found more information, the proposed systems no longer satisfied the new situation.

Bergson pronounced upon this:

> "Because the mind can summarise meaningful moments of time in a single intuition, it frees us from the consternation of a flood of many things. The greater the number of these moments that the mind can combine, the closer is his grip on reality, therefore, the mind of a living being seems to be, firstly, the measure of its power for handling things."

This faculty springs from the fact that man himself is a form of energy. We can see that, initially, creation was explained in a mythological sense. Man spoke about an "upper" and a "lower" world which were personified, and this dualism was explained with a "parental" God from whom both issued, perhaps, in an intuitive sense that energy (stem god) was seen as a synthesis of force and an extension. Putting this into words, classic writers like those of Hindu literature thought apparently along these lines. In the grandiose epic "the dialogue of Ardjuna and Krishna", the blind giant Prakriti wandered unhappy and hungry around the world. He met an even more unhappy, but intelligent, dwarf, Purusha, who suggested that the pair of them would see further when he sat on the shoulder of the giant and they would both be more happy. This seems like a synthesis of force and an extension.

China gave us the eternal values of Lao-Tse, the great inspirer of Emperor Thai-Tsoe, under whose leadership China experienced much growth. Lao-Tse thought that, in the case of man, when force ruled, it led to a boorish way of life. On the other hand, when the mind ruled, it led to frivolity. Really, where force and mind form a synthesis it creates a complete man (see page 77).

Mythology tells us also that the gods themselves are subject to three fundamental elements. The old Greeks combined them into the three Morae (The Fates of the Romans); Clotho spun the thread of life (persistence), Lachesis wove them together (extension), and Atropos cut them with a large pair of scissors (time). In Scandinavia they were called Urda, Verdande and Skuld, the three Norns. The Bhagavad Gitâ named the three Gunas as being the fundamental elements of All, namely:

Satvas Guna	(persistence)
Rajas Guna	(extension)
Tamas Guna	(time)

of whom it was said:

> "Neither on Earth, nor in Heaven, nor under the Gods is there a being that is free of the three Gunas. When Tamas and Rajas are beaten, Satvas has the upperhand, when Rajas and Satvas are hidden, Tamas rules and when Satvas and Tamas are in the minority, Rajas predominates."

Finally from the East came the Three Wise Men who offered gold, myrrh and incense to the new spiritual light. In my opinion this is a tribute which could refer to persistence, extension and time. Perhaps these thoughts on the forms of energy, which have roots amongst all people can contribute to our world in evolution where the main stages are: energy — matter — vitalisation of life — and the humanisation of life. To the extension of this last stage could be assigned the dimension L^{20}.

TABLE 7

	Energy	*Force*	*Extension*
Mythology			
Greece	Chronos	Uranus	Gaia
Babylon	Apsoe	Eshara	Noedimaroed
Hinduism	Brahma	Shiwa	Vishnoe
Scandinavian	Wodan	Balder	Loki
Shinto	Kuni-toko-tachi	Izanami	Izanagi
Aztecs	Huitzilo-pachtli	Quetzalcoatl	Tlaloc
Theology			
Old testament	Paradise	Tree of knowledge	Tree of life
Bhagavadgita	Brahma	Purusha	Prakriti
Tau-te-tjing	Tau	Yang	Yin
Koran	Allah	Plan	Implementation
Philosophy			
Heraclitus	Fire	Being	Becoming
Spinoza	Substance	Natura naturata	Natura naturans
Hegel	Synthesis	Thesis	Antithesis
Bergson	Energy	Memory	Action

Appendix 1

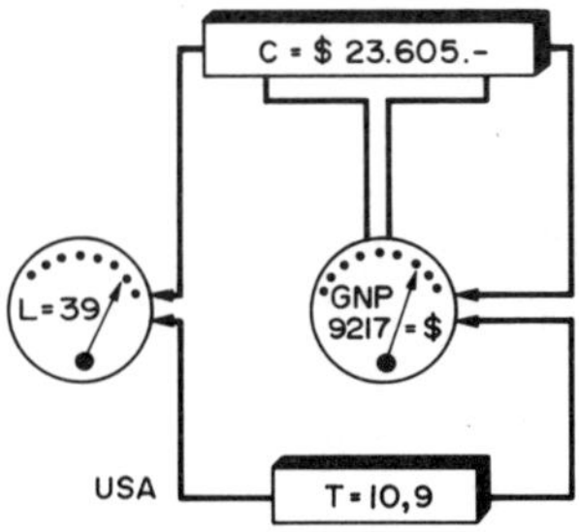

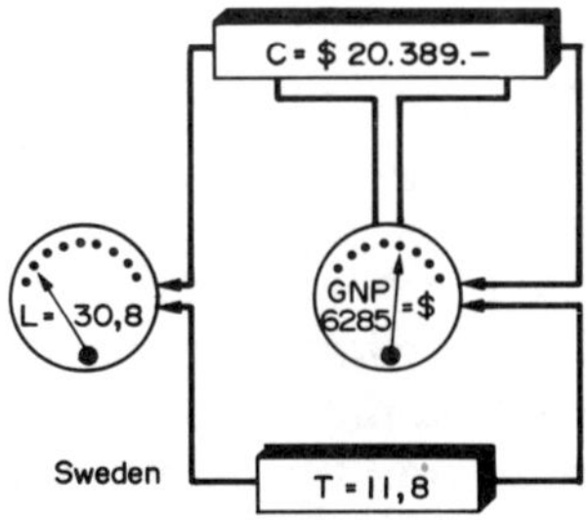

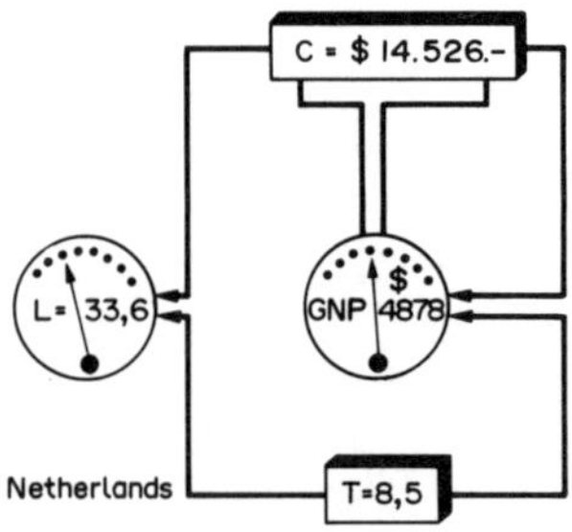

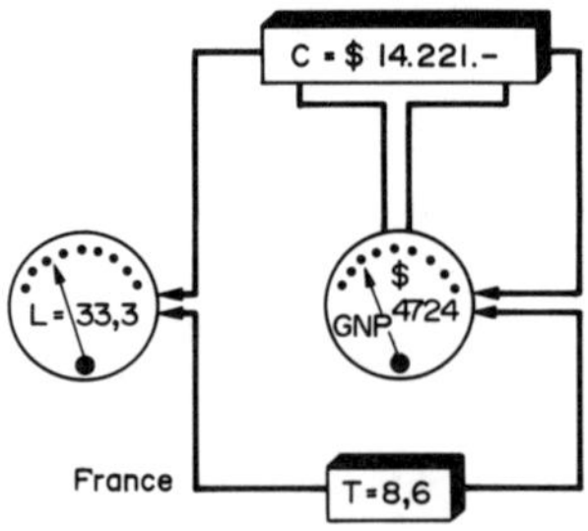

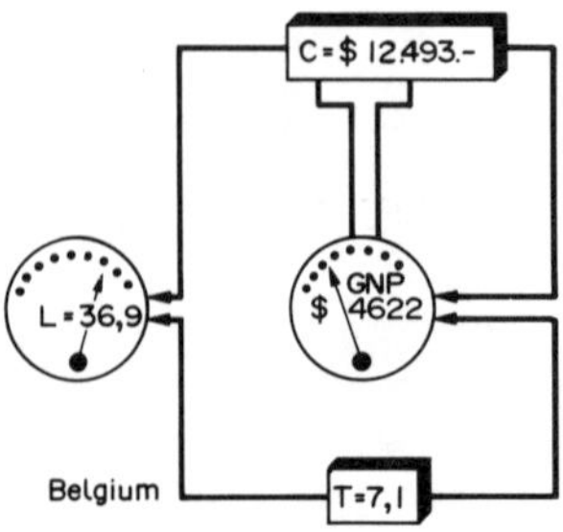

Macro - economics

GNP: Gross national product
C: Capital investment
L: Laboriousness
T: Technological qualification

TABLE 8

Hierarchy System of Energies E = Force . Extension $E_n = (M.L^{2-n}.T^{-2}).(L^n)$
$n = 1, 2, 3, \ldots$

Dimension	Time	Persistence	Extension	Action	Impulse	Force	Power	Energy
System	[t]	[p]	[e]	$[v]=[et^{-1}]$	$[pv]=[pet^{-1}]$	$[F]=[pvt^{-1}]=[pet^{-2}]$	$[P]=[Fv]=[pe^2t^{-3}]$	$[E]=[Pt]=[pe^2t^{-2}]$
Macro Economy	T	ML^{-36}	L^{19}	Productive actions	Impulses	Cap.goods	G.N.P.	Energy
3D Info-Syst	T	Control		Managerial actions	Impulses	Planning syst.	Power	Energy
2D Info-Syst	T	Control		Managerial actions	Impulses	Planning syst.	Power	Energy
1D Info-Syst	T	ML^{-30}	L^{16}	Managerial actions	Impulses	Planning syst.	Power	Energy
Organisation	T	Values		Interactions	Impulses	Ordering	Power	Energy
Group	T	ML^{-26}	L^{14}	Interactions	Impulses	Norm-setting	Power	Energy
Homo Faber	T	Perception	Motivation	Actions	Impulses	Ability	Power	Energy
Cerebral Cortex	T	Sense organ		Info-transport	Impulses	Memory	Power	Energy
Ganglion		ML^{-20}	L^{11}	Info-transport	Impulses	Node	Power	Energy
Neuron	T	Receptor cell		Info-transport	Impulses	Node	Power	Energy
Organism	T	ML^{-16}	L^9	Stimuli	Impulses	DNA (coded experience)	Power	Energy
Tissue	T	Receptor Membrane		Stimuli	Impulses	DNA (coded experience)	Power	Energy
Cell	T	ML^{-12}	L^7	Stimuli	Impulses	DNA (coded experience)	Power	Energy
Crystal	T	ML^{-10}	L^6	L^6T^{-1}	$ML^{-4}T^{-1}$	EL^{-6}	ML^2T^{-3}	ML^2T^{-2}
Volume Elasticity	T						ML^2T^{-3}	ML^2T^{-2}
Torsion	T						ML^2T^{-3}	ML^2T^{-2}
Hydraulics	T	ML^{-4}	L^3	L^3T^{-1}	$ML^{-1}T^{-1}$	EL^{-3}	ML^2T^{-3}	ML^2T^{-2}
Areal Energy	T	.					ML^2T^{-3}	ML^2T^{-2}
Translation	T	M	L^1	L^1T^{-1}	MLT^{-1}	EL^{-1}	ML^2T^{-3}	ML^2T^{-2}
Rotation	T	ML^2	L^0	L^0T^{-1}	ML^2T^{-1}	E	ML^2T^{-3}	ML^2T^{-2}
Electro Magn.	T	ML^4	L^{-1}	$L^{-1}T^{-1}$	ML^3T^{-1}	EL^1	ML^2T^{-3}	ML^2T^{-2}
Nuclear Energy	T						ML^2T^{-3}	ML^2T^{-2}
Chromo Dynamical Energy	T	ML^8	L^{-3}	$L^{-3}T^{-1}$	ML^5T^{-1}	EL^3	ML^2T^{-3}	ML^2T^{-2}
Diffraction	T						ML^2T^{-3}	ML^2T^{-2}
Pulsing of Universe	T						ML^2T^{-3}	ML^2T^{-2}
Primeval Atom	T	ML^{14}	L^{-6}	$L^{-6}T^{-1}$	ML^8T^{-1}	EL^6	ML^2T^{-3}	ML^2T^{-2}

System \ Dimension	Persistence	Extension	Force	Sub systems (page 81)	Species
Macro econ.	Know-how	Labour	Tools	8 Economic systems	
3D info. syst.	Control	Implement	Integr. plan. syst.	8 3D-production systems	
2D info. syst.				8 2D-production systems	
ID info. syst.				8 goods exchange systems	
Organisation	Values	Inter. actions	Norm setting	8 types of organisation	
Group				8 types of groups	
Homo-faber	Perception	Action	Ability	8 psychychological types	
Cortex	Sense organ	Info. transport	Memory	8 classes 3-germlayers	
Ganglion				8 classes 2-germlayers	
Neuron	Receptor cell	Info. transport	Node	8 classes protozoa	
Organism				8 classes cormophyta	
Tissue				8 classes thallophyta	
Cell	Receptor membrane	Stimuli	Reg. gene	8 classes protophyta	
Crystal	ML^{-10}	L^6	$ML^{-4}T^{-2}$	8 crystalline systems	
Vol. elast.					
Torsion					
Hydraulics	ML^{-4}	L^3	$ML^{-1}T^{-2}$	8 hydraulic systems	
Areal-energy					
Translation	M	L	MLT^{-2}	8 mechanical systems	
Rotation	ML^2	α	ML^2T^{-2}		
Electr. magn				8 fundamental periods	
Nucl. energy				Fundamental particles	
Chromo dyn. en				Quarks	
Diffraction					
Pulsation					
Primeval atom					

$$E_n = \langle M.L^{2-n}.T^{-2}\rangle.\langle L^n\rangle$$

n = 1, 2, 3..........

Energetica-sub system

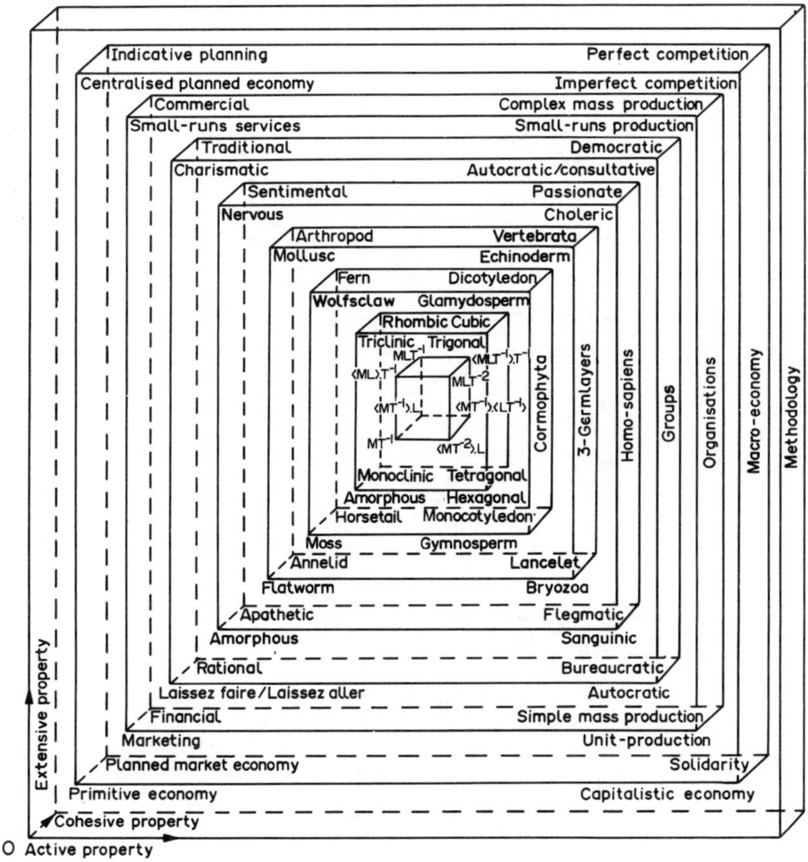

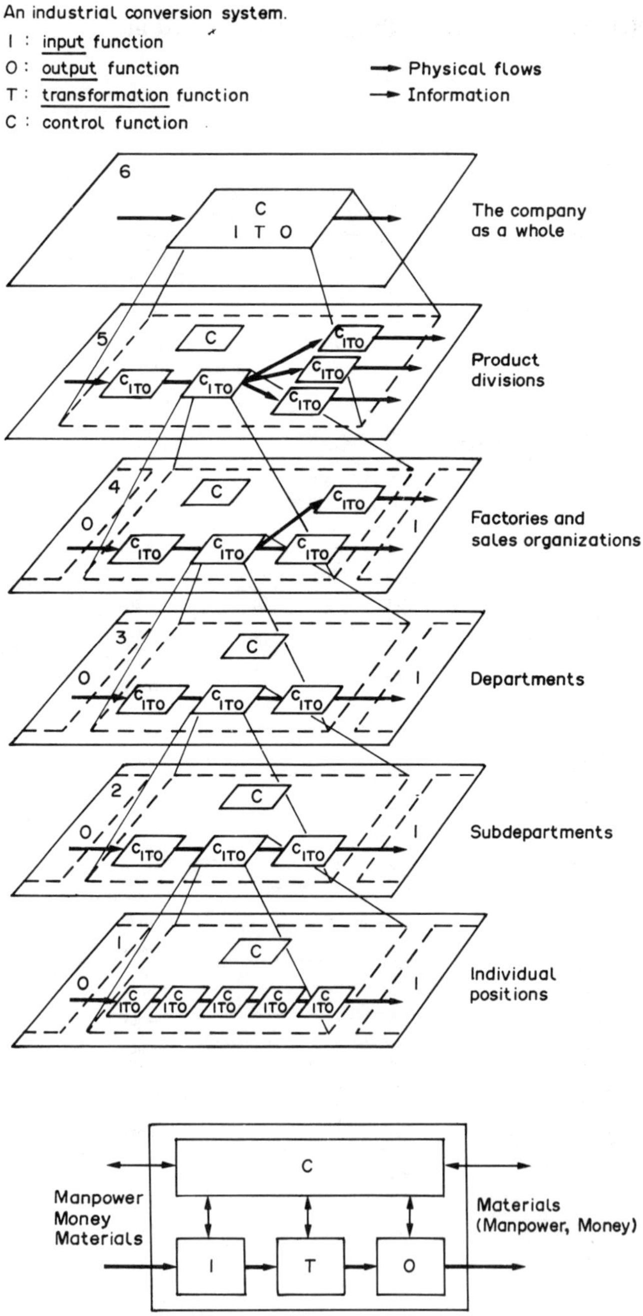

An industrial conversion system.
I : input function
O : output function
T : transformation function
C : control function
Physical flows
Information
6
C
I T O
The company as a whole
5
Product divisions
4
Factories and sales organizations
3
Departments
2
Subdepartments
1
Individual positions
C
I
T
O
Manpower
Money
Materials
Materials
(Manpower, Money)

PDP 11/40 Computer Program for a speed and skilfulness test

```
                            .TITLE  GVPSY2
                            .MCALL  .REGDEF,.EXIT,.PRINT,.QSET,.MRKT,.TTYIN
                            .MCALL  .CMKT,.SPND,.RSUM,..V2..
000000                      ..V2..
000000                      .REGDEF
        000177              RO=177
        000040              BL=40
                            .GLOBL  INTOUT,LPSLED

00000   005067 START:       CLR     LED
        000464
00004                       .QSET   #QAREA,#1         ;FOR MRKT
00016                       .LNKRT                    ;LINK TO MONITOR
00022                       .SCROL  #SCBUF            ;SET UP SCROLLER
00032                       .INSRT  #DFILE            ;START DFILE
00042                       .LPEN   #LPBUF            ;SET UP LPEN
00052   005267              INC     LPBUF             ;DISABLE LPEN HITS
        000502
00056                       .TTYIN                    ;WAIT FOR START COMM.
00062   005002              CLR     R2                ;POINTER IN ANSWER TABLE
00064   012767              MOV     #300.,SEC         ;DELAY,6 SEC.
        000454
        000426
00072   012703              MOV     #TEL,R3
        000576'
00076   012304 L1:          MOV     (R3)+,R4          ;# OF DISP. FOR TRIAL
00100   012701              MOV     #900.,R1
        001604
00104   010167              MOV     R1,DX
        001062
00110   052767              BIS     #INTX,DX
        040000
        001054
00116   005000              CLR     R0
00120   071004              DIV     R4,R0
00122   010067              MOV     R0,DELTA
        001032
00126   005005              CLR     R5                ;CLEAR SCORE
00130   005267 L2:          INC     LED
        000334
00134   004567              JSR     R5,LPSLED
        000000G
00140   000470'             LED
00142   062767              ADD     #24.,ENTRY        ;UPDATE DISPLAY
        000030
        001116
00150   005067              CLR     LPBUF             ;ENABLE LPEN HIT
        000404
00154                       .MRKT   #MTAREA,#TIME,#MTCMPL,#1
00214   005067              CLR     MTFLAG
        000252
00220   005067              CLR     HIT
        000262
00224   005767 L4:          TST     HIT
        000256

00230   001003              BNE     1$
00232   012767              MOV     #NOCHAR,SUBR
        001302'
        001034
00240   005767 1$:          TST     LPBUF             ;LPEN HIT?
        000314
00244   001423              BEQ     L3                ;NO,BRANCH
00246   005267              INC     HIT
        000234
00252   016700              MOV     LPBUF+6,R0        ;GET DPC
        000310
00256   162700              SUB     #TARGET,R0        ;IN TARGET AREA?
        001204'
00262   100414              BMI     L3                ;NO,BRANCH
00264   020027              CMP     R0,#46            ;STILL IN TARGET AREA?
        000046
00270   003011              BGT     L3                ;NO,BRANCH
00272   006200              ASR     R0
00274   006200              ASR     R0
00276   006200              ASR     R0                ;CALCULATE OFFSET
00300   110067              MOVB    R0,ANSWER
        000204
00304   006300              ASL     R0
00306   016067              MOV     TABLE(R0),SUBR
        000474'
        000760
00314   005067 L3:          CLR     LPBUF             ;DISABLE LPEN HITS
        000240
00320   005767              TST     MTFLAG
        000146
00324   001001              BNE     L6
00326   000736              BR      L4

00330   126762 L6:          CMPB    ANSWER,GOOD(R2)
        000154
        000650'
00336   001001              BNE     2$
00340   005205              INC     R5
00342   005202 2$:          INC     R2                ;INCR. POINTER IN ANSWER
```

```
TABLE
00344 032737         BIT     #1,@#177570
      000001
      177570
00352 001040         BNE     EXIT
00354 005304         DEC     R4
00356 001413         BEQ     1$
00360 042767         BIC     #INTX,DX
      040000
      000604
00366 166767         SUB     DELTA,DX
      000566
      000576
00374 052767         BIS     #INTX,DX
      040000
      000570
00402 000167         JMP     L2
      177522
00406 105267 1$:     INCB    TRIAL
      000206
00412 010567         MOV     R5,L5
      000006
00416 004567         JSR     R5,INTOUT
      000000G
00422    014         .BYTE   12.
00423    240         .BYTE   240
00424 000000 L5:     0
00426 000632'        BUFFER
00430                .PRINT  #MSG
00436 162767         SUB     #50.,SEC          ;CALCULATE NEW INTERVAL
      000062
      000054
00444 026727         CMP     SEC,#100.         ;DONE?
      000050
      000144
00452 001211         BNE     L1                ;NO,TRY AGAIN
00454        EXIT:   .UNLNK                    ;UNLINK
00460                .EXIT                     ;EXIT

00462 005267 MTCMPL: INC     MTFLAG
      000004
00466 000207         RTS     PC
00470 000000 LED:    0
00472 000000 MTFLAG: 0

00474 001306'TABLE:  CHAR0
00476 001324'        CHAR1
00500 001342'        CHAR2
00502 001360'        CHAR3
00504 001376'        CHAR4

00506 000000 HIT:    0
00510 000000 ANSWER: 0
00512    005 SCBUF:  .BYTE   5,0
00513    000
0514 000144          100.
0516 000000 TIME:    0
0520 000454 SEC:     6*50.
0522        MTAREA:  .BLKW   4
0532        MAREA1:  .BLKW   4
0542        QAREA:   .BLKW   7
0560        LPBUF:   .BLKW   7
0576 000036 TEL:     30.,36.,45.,60.
0600 000044
0602 000055
0604 000074
0606    122 MSG:     .ASCII  /RESULTAAT /
0607    105
0610    123
0611    125
0612    114
0613    124
0614    101
0615    101
0616    124
0617    040
0620    060 TRIAL:   .BYTE   '0,'E
0621    105
0622    040          .ASCII  / TRIAL :/
0623    124
0624    122
0625    111
0626    101
0627    114
0630    040
0631    072
0632        BUFFER:  .BLKW   6
0646 000000          0
                     .NLIST  BEX
0650    004 GOOD:    .BYTE   4,4,4,1,0,1,1,4,4,1
0662    001          .BYTE   1,4,2,4,4,1,4,4,2,3
0674    003          .BYTE   3,0,3,4,4,3,0,1,1,0
0706    004          .BYTE   4,2,1,3,1,1,4,4,4,0
0720    001          .BYTE   1,4,3,1,2,4,2,3,1,4
```

```
0732    000          .BYTE   0,0,2,3,1,1,0,4,2,0
0744    001          .BYTE   1,1,2,1,2,1,3,0,2,1
0756    002          .BYTE   2,0,0,4,3,0,0,3,3,1
0770    003          .BYTE   3,0,0,3,0,0,0,2,3,1
1002    003          .BYTE   3,4,3,0,3,4,2,0,4,3
1014    003          .BYTE   3,0,0,2,3,0,1,3,0,2
1026    001          .BYTE   1,3,3,0,3,0,0,2,0,1
1040    003          .BYTE   3,3,2,3,2,1,4,4,1,1
1052    001          .BYTE   1,3,0,1,4,0,0,0,3,2
1064    000          .BYTE   0,4,3,1,4,3,4,0,2,2
1076    002          .BYTE   2,0,0,0,1,4,3,0,4,0
1110    000          .BYTE   0,4,4,4,0,2,1,1,4,4
1122    004          .BYTE   4,2,1,3,0,1,4,3,0,4
1134    004          .BYTE   4,2,0,1,1,2,0,2,2,1
1146    002          .BYTE   2,1,0,1,3,2,3,2,1,2

1160 000000 DELTA:   0
1162 114000 DFILE:   POINT,0,700.
1170 113120          LONGV!INT4!BLKOFF!LPOFF
1172 041604 DX       900.!INTX
1174 000000          0
1176 114000          POINT,400.,310.
1204 107560 TARGET:  SHORTV!INT6!BLKOFF!LPON
1206 047000          28.*200!INTX
1210 132120          RELATV!INT0!BLKOFF!LPOFF
1212 007000          28.*200
1214 107560          SHORTV!INT6!BLKOFF!LPON
1216 047000          28.*200!INTX
1220 132120          RELATV!INT0!BLKOFF!LPOFF
1222 007000          28.*200
1224 107560          SHORTV!INT6!BLKOFF!LPON
1226 047000          28.*200!INTX
1230 132120          RELATV!INT0!BLKOFF!LPOFF
1232 007000          28.*200
1234 107560          SHORTV!INT6!BLKOFF!LPON
1236 047000          28.*200!INTX
1240 132120          RELATV!INT0!BLKOFF!LPOFF
1242 007000          28.*200
1244 107560          SHORTV!INT6!BLKOFF!LPON
1246 047000          28.*200!INTX
1250 132120          RELATV!INT0!BLKOFF!LPOFF
1252 007000          28.*200
1254 114000          POINT,400.,350.
1262 173400          DJSR
1264 001270'         .+4
1266 001414'ENTRY:   SUB0
1270 173400          DJSR
1272 001276'         .+4
1274 001302'SUBR:    NOCHAR
1276 173400          DRET
1300 000000          0

     000606          Y0=390.
     000627          X0=407.
     000717          X1=X0+56.
     001007          X2=X1+56.
     001077          X3=X2+56.
     001167          X4=X3+56.
1302 173400 NOCHAR:  DRET,0
1306 114000 CHAR0:   POINT,X0,Y0
1314 103130          CHAR!INT4!BLKON!LPOFF
1316    052          .ASCII  /*/
1320 173400          DRET,0
1324 114000 CHAR1:   POINT,X1,Y0
1332 103130          CHAR!INT4!BLKON!LPOFF
1334    052          .ASCII  /*/
1336 173400          DRET,0
1342 114000 CHAR2:   POINT,X2,Y0
1350 103130          CHAR!INT4!BLKON!LPOFF
1352    052          .ASCII  /*/
1354 173400          DRET,0
1360 114000 CHAR3:   POINT,X3,Y0
1366 103130          CHAR!INT4!BLKON!LPOFF
1370    052          .ASCII  /*/
1372 173400          DRET,0
1376 114000 CHAR4:   POINT,X4,Y0
1404 103130          CHAR!INT4!BLKON!LPOFF
1406    052          .ASCII  /*/
1410 173400          DRET,0

1414 103120 SUB0:    CHAR!INT4!BLKOFF!LPOFF
```

References

Chapter 1

BACHELARD, G. *Le materialisme rationnel.* Paris Presses, Universitaires de France, 1953.

BRONOWSKI, J. New Concepts in the Evolution of Complexity: Stratified stability and unbounded Plans, *Zygon* **vi**, March 1970.

LASZLO, E. *The Systems View of the World.* New York, Braziller, 1972.

MILLER, J. G. *Living Systems.* New York, McGraw–Hill 1978.

OSTWALD, W. *Energetische Grundlagen der Kulturwissenschaften.* Leipzig, Klinkhardt, 1909.

WHYTE, L. L. *The Universe of Experience.* New York, Harper & Row, 1962.

Chapter 2

ALBERDA, Th. *De groene aarde: biologie en ecologie van de plant.* Utrecht, Spectrum, 1966.

FAST, J. D. *Materie en leven.* Heerlen, Centrale Uitgeverij en adviesbureau, 1972.

Chapter 3

HEISENBERG, N. *Physics and Philosophy,* New York, Harper & Row, 1962.

KARNOPP, D. and ROSENBERG, R. *Systems Dynamics, a Unified Approach.* East Lansing, Mich., MSU, 1972.

MARGENAU, H. *The Nature of Physical Reality.* New York, McGraw-Hill, 1950.

PAYNTER, H. M. *Analysis and Design of Engineering Systems.* Cambridge, Mass., MIT Press, 1961.

POLDER, J. W. *Begrippen, Symbolen, Eenheden.* Eindhoven, Eindhoven University of Technology, 1974.

THOMA, J. Bond Graphs for Thermal Energy Transport and Entropy Flow, *J. Franklin Institute,* **292** 109–20.

Chapter 4

BUCKLEY, W. *Modern Systems Research for the Behavioral Scientist.* Chicago, Aldine, 1968.

DOORN, J. A. A. van and LAMMERS, C. J. *Moderne Sociologie.* Utrecht, Spectrum, 1962.

KAHN, H. and WIENER, A. J. *The Year 2000.* New York, Macmillan, 1968.

KRUPP, S. R. *The Structure of Economic Science.* Englewood Cliffs, NJ., Prentice-Hall, 1966.

Chapter 5

BUCKLEY, W. *Modern Systems Research for the Behavioral Scientist.* Chicago, Aldine, 1968.

ENGLEBARI, D. C. *Study for the Development of Human Intellect Augmentation Techniques.* SRI project 5890, 1967 (restricted).
KORMAN, A. K. *Industrial and Organizational Psychology.* Englewood Cliffs, NJ, Prentice-Hall, 1971.
LEAVITT, H. J. *Managerial Psychology.* Chicago, University of Chicago Press, 1972.
LIKERT, R. *New Patterns of Management.* New York, McGraw-Hill, 1961.
LINDSLEY, D. B. Psychophysiology and Motivation. In *Nebraska Symp. on Motivation 1957*, M. R. Jones, ed. Lincoln, University of Nebraska Press, 1957.
MASLOW, A. H. *Motivation and Personality.* New York, Harper & Row, 1954.
McGREGOR, D. *The Human Side of Enterprise.* New York, McGraw-Hill, 1960.
VROOM, V. H. *Work and Motivation.* New York, Wiley, 1967.

Chapter 6

BLAKE, R. R. and MOUTON, J. S. *Group Dynamics: Key to Decision Making.* Houston, Gulf, 1961.
COLTHOF, H. *Veranderen en aanpassen in de organisatie van het bedrijf.* Alphen aan den Rijn, Samsom, 1965.
MERTON, R. K. *Social Theory and Social Structure.* New York, Free Press, 1968.
MITCHAM, C. and MACKY, R. *Philosophy and Technology.* New York, Free Press, 1972.
ZUTHEM, H. J. VAN *Arbeid en Arbeidsbeleid in de Onderneming.* Assen, Van Gorcum, 1971.

Chapter 7

ANSOFF, H. I. The Expanding Role of the Computer in Managerial Decision Making, *Informatie,* **9**, 46–64.
BOYCE, R. O. *Integrated Managerial Controls,* London, Longmans, 1967.
BRECH, E. F. L. *Organisation; the Framework of Management.* London, Longmans & Green, 1958.
BROWN, R. G. *Smoothing, Forecasting and Prediction of Discrete Time Series.* London, Prentice-Hall, 1963.
FORRESTER, J. W. *Industrial Dynamics.* Cambridge, Mass., MIT Press, 1965.
GEORGE, C. S. *The History of Management Thought.* Englewood Cliffs, N.J., Prentice-Hall, 1968.
JANTSCH, F. *Perspectives of Planning.* Paris, OECD, 1969.
JOHNSON, R. A., KAST, F. E. and ROSENZWEIG, J. E. *The Theory and Management of Systems.* New York, McGraw-Hill, 1967.
MARCH, J. G. and SIMON, H. A. *Organizations.* New York, Wiley, 1961.
NIELEN, G. C. J. F. *Informatiesystemen en het besturen van ondernemingen.* Alphen aan den Rijn, Samsom. 1969.
SIMON, H. A. *Administrative Behaviour.* New York, Macmillan, 1958.

Chapter 8

BOMBACH, G. U. A. (Hrsg.) *Wachstum und Konjunktur.* Darmstadt, Leske, 1960.
BROOMAN, F. S. *Macroeconomics.* London, Allen & Unwin, 1963.
CORREA, H. *The Economics of Human Resources.* Den Haag, Pasmans, 1962.
MYRDAL, G. *Economic Theory and Under-developed Regions.* London, Methuen, 1963.
PEN, J. De neo-klassieke theorie en de technische vooruitgang, *De Economist,* **115**, 39–59.
ROBERTS, R. S. *Economic Development, Human Skills and Technical Assistance.* Geneva, Droz, 1960.
SOLOW, R. M. *Capital Theory and the Rate of Return.* Amsterdam, North-Holland, 1964.
TINBERGEN, J. *Selected Papers.* Amsterdam, North-Holland, 1959.

Chapter 11

Bergson, H. Matière et mémoire. In *Oeuvres: essai sur les données immediates de la conscience.* Paris, Presses universitaires de France, 1959.
Bosch, E. D. K. *De gouden kiem.* Amsterdam, Elsevier, 1948.
Fung Yu-Lan. *A History of Chinese Philosophy*. Princeton, N.J., Princeton University Press, 1952.
Kuhn, T. S. *The Structure of Scientific Revolutions.* Chicago, University of Chicago Press, 1970.
Macnicol, N. *Hindu Scriptures.* London, Everyman's Library, 1938.
Teilhard de Chardin, P. *Le phénomène humain.* Paris, Editions du Seuil, 1955.

Name Index

Subject Index

C.P.C. ABERYSTWYTH LLYFRGELL